U0138285

TOUCHSTONE

剑桥标准英语教程

MICHAEL MCCARTHY
JEANNE MCCARTEN
HELEN SANDIFORD

1

STUDENT'S BOOK
学生用书

CAMBRIDGE
UNIVERSITY PRESS

北京语言大学出版社
BEIJING LANGUAGE AND CULTURE
UNIVERSITY PRESS

图书在版编目(CIP)数据

剑桥标准英语教程1学生用书／(英)麦卡锡
(McCarthy，M.)，(英)麦克卡顿(McCarten，J.)，(英)
桑迪福德(Sandiford，H.)编著. —北京：北京语言大
学出版社，2010 (2011.7 重印)
ISBN 978-7-5619-2649-9

Ⅰ.①剑…　Ⅱ.①麦…②麦…③桑…　Ⅲ.①英语—
教材　Ⅳ.①H31

中国版本图书馆 CIP 数据核字（2010）第 015436 号

版权登记：图字 01－2009－7377 号

This is a reprint edition of the following title published by Cambridge University Press：

ISBN 978-0-521-66611-4 Touchstone Student's Book 1 with Audio CD/CD-ROM

ⓒ Cambridge University Press 2005

This reprint edition for the People's Republic of China (excluding Hong Kong，Macau and Taiwan) is published by arrangement with the Press Syndicate of the University of Cambridge，Cambridge，United Kingdom.

ⓒ Cambridge University Press and Beijing Language and Culture University Press 2010

书　　　名：剑桥标准英语教程1·学生用书
编　　　著：(英) Michael McCarthy，(英) Jeanne McCarten，(英) Helen Sandiford
责任编辑：余心乐　刘格均
封面设计：大愚设计 ＋ 赵文康

出版发行：**北京语言大学出版社**
社　　　址：北京市海淀区学院路 15 号　邮政编码：100083
网　　　站：www. blcup. com
电　　　话：发行部　(010)62605588 /5019 /5128
　　　　　　编辑部　(010)62605189
　　　　　　邮购电话　(010)62605127
　　　　　　读者服务信箱　bj62605588@163.com
印　　　刷：北京画中画印刷有限公司
经　　　销：全国新华书店
版　　　次：2011 年 7 月第 1 版第 2 次印刷
开　　　本：889 毫米×1194 毫米　1/16　印张：9.75
字　　　数：198 千
书　　　号：ISBN 978-7-5619-2649-9
定　　　价：59.00 元

Authors' acknowledgments

Touchstone has benefited from extensive development research. The authors and publishers would like to extend their particular thanks to the following reviewers, consultants, and piloters for their valuable insights and suggestions.

Reviewers and consultants:

Thomas Job Lane and Marilia de M. Zanella from **Associação Alumni**, São Paulo, Brazil; Simon Banha from **Phil Young's English School**, Curitiba, Brazil; Katy Cox from **Casa Thomas Jefferson**, Brasilia, Brazil; Rodrigo Santana from **CCBEU**, Goiânia, Brazil; Cristina Asperti, Nancy H. Lake, and Airton Pretini Junior from **CEL LEP**, São Paulo, Brazil; Sonia Cury from **Centro Britânico**, São Paulo, Brazil; Daniela Alves Meyer from **IBEU**, Rio de Janeiro, Brazil; Ayeska Farias from **Mai English**, Belo Horizonte, Brazil; Solange Cassiolato from **LTC**, São Paulo, Brazil; Fernando Prestes Maia from **Polidiomas**, São Paulo, Brazil; Chris Ritchie and Debora Schisler from **Seven Idiomas**, São Paulo, Brazil; Maria Teresa Maiztegui and Joacyr de Oliveira from **União Cultural EEUU**, São Paulo, Brazil; Sakae Onoda from **Chiba University of Commerce**, Ichikawa, Japan; James Boyd and Ann Conlon from **ECC Foreign Language Institute**, Osaka, Japan; Catherine Chamier from **ELEC**, Tokyo, Japan; Janaka Williams, Japan; David Aline from **Kanagawa University**, Yokohama, Japan; Brian Long from **Kyoto University of Foreign Studies**, Kyoto, Japan; Alistair Home and Brian Quinn from **Kyushu University**, Fukuoka, Japan; Rafael Dovale from **Matsushita Electric Industrial Co., Ltd.**, Osaka, Japan; Bill Acton, Michael Herriman, Bruce Monk, and Alan Thomson from **Nagoya University of Commerce**, Nisshin, Japan; Alan Bessette from **Poole Gakuin University**, Osaka, Japan; Brian Collins from **Sundai Foreign Language Institute, Tokyo College of Music**, Tokyo, Japan; Todd Odgers from **The Tokyo Center for Language and Culture**, Tokyo, Japan; Jion Hanagata from **Tokyo Foreign Language College**, Tokyo, Japan; Peter Collins and Charlene Mills from **Tokai University**, Hiratsuka, Japan; David Stewart from **Tokyo Institute of Technology**, Tokyo, Japan; Alberto Peto Villalobos from **Cenlex Santo Tomás**, Mexico City, Mexico; Diana Jones and Carlos Lizarraga from **Instituto Angloamericano**, Mexico City, Mexico; Raúl Mar and María Teresa Monroy from **Universidad de Cuautitlán Izcalli**, Mexico City, Mexico; JoAnn Miller from **Universidad del Valle de México**, Mexico City, Mexico; Orlando Carranza from **ICPNA**, Peru; Sister Melanie Bair and Jihyeon Jeon from **The Catholic University of Korea**, Seoul, South Korea; Peter E. Nelson from **Chung-Ang University**, Seoul, South Korea; Joseph Schouweiler from **Dongguk University**, Seoul, South Korea; Michael Brazil and Sean Witty from **Gwangwoon University**, Seoul, South Korea; Kelly Martin and Larry Michienzi from **Hankook FLS University**, Seoul, South Korea; Scott Duerstock and Jane Miller from **Konkuk University**, Seoul, South Korea; Athena Pichay from **Korea University**, Seoul, South Korea; Lane Darnell Bahl, Susan Caesar, and Aaron Hughes from **Korea University**, Seoul, South Korea; Farzana Hyland and Stephen van Vlack from **Sookmyung Women's University**, Seoul, South Korea; Hae-Young Kim, Terry Nelson, and Ron Schafrick from **Sungkyunkwan University**, Seoul, South Korea; Mary Chen and Michelle S. M. Fan from **Chinese Cultural University**, Taipei, Taiwan, China; Joseph Sorell from **Christ's College**, Taipei, Taiwan, China; Dan Aldridge and Brian Kleinsmith from **ELSI**, Taipei, Taiwan, China; Ching-Shyang Anna Chien and Duen-Yeh Charles Chang from **Hsin Wu Institute of Technology**, Taipei, Taiwan, China; Timothy Hogan, Andrew Rooney, and Dawn Young from **Language Training and Testing Center**, Taipei, Taiwan, China; Jen Mei Hsu and Yu-hwei Eunice Shih from **Taiwan Normal University**, Taipei, Taiwan, China; Roma Starczewska and Su-Wei Wang from **PQ3R Taipei Language and Computer Center**, Taipei, Taiwan, China; Elaine Parris from **Shih Chien University**, Taipei, Taiwan, China; Jennifer Castello from **Cañada College**, Redwood City, California, USA; Dennis Johnson, Gregory Keech, and Penny Larson from **City College of San Francisco – Institute for International Students**, San Francisco, California, USA; Ditra Henry from **College of Lake County**, Gray's Lake, Illinois, USA; Madeleine Murphy from **College of San Mateo**, San Mateo, California, USA; Ben Yoder from **Harper College**, Palatine, Illinois, USA; Christine Aguila, John Lanier, Armando Mata, and Ellen Sellergren from **Lakeview Learning Center**, Chicago, Illinois, USA; Ellen Gomez from **Laney College**, Oakland, California, USA; Brian White from **Northeastern Illinois University**, Chicago, Illinois, USA; Randi Reppen from **Northern Arizona University**, Flagstaff, Arizona, USA; Janine Gluud from **San Francisco State University – College of Extended Learning**, San Francisco, California, USA; Peg Sarosy from **San Francisco State University – American Language Institute**, San Francisco, California, USA; David Mitchell from **UC Berkley Extension, ELP – English Language Program**, San Francisco, California, USA; Eileen Censotti, Kim Knutson, Dave Onufrock, Marnie Ramker, and Jerry Stanfield from **University of Illinois at Chicago – Tutorium in Intensive English**, Chicago, Illinois, USA; Johnnie Johnson Hafernik from **University of San Francisco, ESL Program**, San Francisco, California, USA; Judy Friedman from **New York Institute of Technology**, New York, New York, USA; Sheila Hackner from **St. John's University**, New York, New York, USA; Joan Lesikin from **William Paterson University**, Wayne, New Jersey, USA; Linda Pelc from **LaGuardia Community College**, Long Island City, New York, USA; Tamara Plotnick from **Pace University**, New York, USA; Lenore Rosenbluth from **Montclair State University**, Montclair, New Jersey, USA; Suzanne Seidel from **Nassau Community College**, Garden City, New York, USA; Debbie Un from **New York University, New School**, and **LaGuardia Community College**, New York, New York, USA; Cynthia Wiseman from **Hunter College**, New York, New York, USA; Aaron Lawson from **Cornell University**, Ithaca, New York, USA, for his help in corpus research; Belkis Yanes from **CTC Belo Monte**, Caracas, Venezuela; Victoria García from **English World**, Caracas, Venezuela; Kevin Bandy from **LT Language Teaching Services**, Caracas, Venezuela; Ivonne Quintero from **PDVSA**, Caracas, Venezuela.

Piloters:

Daniela Jorge from **ELFE Idiomas**, São Paulo, Brazil; Eloisa Marchesi Oliveira from **ETE Professor Camargo Aranha**, São Paulo, Brazil; Marilena Wanderley Pessoa from **IBEU**, Rio de Janeiro, Brazil; Marcia Lotaif from **LTC**, São Paulo, Brazil; Mirlei Valenzi from **USP English on Campus**, São Paulo, Brazil; Jelena Johanovic from **YEP International**, São Paulo, Brazil; James Steinman from **Osaka International College for Women**, Moriguchi, Japan; Brad Visgatis from **Osaka International University for Women**, Moriguchi, Japan; William Figoni from **Osaka Institute of Technology**, Osaka, Japan; Terry O'Brien from **Otani Women's University**, Tondabayashi, Japan; Gregory Kennerly from **YMCA Language Center** piloted at **Hankyu SHS**, Osaka, Japan; Daniel Alejandro Ramos and Salvador Enríquez Castaneda from **Instituto Cultural Mexicano-Norteamericano de Jalisco**, Guadalajara, Mexico; Patricia Robinson and Melida Valdes from **Universidad de Guadalajara**, Guadalajara, Mexico.

We would also like to thank the people who arranged recordings: Debbie Berktold, Bobbie Gore, Bill Kohler, Aaron Lawson, Terri Massin, Traci Suiter, Bryan Swan, and the many people who agreed to be recorded.

The authors would also like to thank the **editorial** and **production** team:
Sue Aldcorn, Eleanor K. Barnes, Janet Battiste, Sylvia P. Bloch, David Bohlke, Karen Brock, Jeff Chen, Sylvia Dare, Karen Davy, Deborah Goldblatt, Paul Heacock, Louisa Hellegers, Eliza Jensen, Lesley Koustaff, Heather McCarron, Lise R. Minovitz, Diana Nam, Kathy Niemczyk, Bill Paulk, Bill Preston, Janet Raskin, Mary Sandre, Tamar Savir, Shelagh Speers, Kayo Taguchi, Mary Vaughn, Jennifer Wilkin, and all the design and production team at Adventure House.

And these Cambridge University Press **staff** and **advisors**:
Yumiko Akeba, Jim Anderson, Kanako Aoki, Mary Louise Baez, Carlos Barbisan, Alexandre Canizares, Cruz Castro, Kathleen Corley, Kate Cory-Wright, Riitta da Costa, Peter Davison, Elizabeth Fuzikava, Steven Golden, Yuri Hara, Catherine Higham, Gareth Knight, João Madureira, Andy Martin, Alejandro Martínez, Nigel McQuitty, Carine Mitchell, Mark O'Neil, Rebecca Ou, Antonio Puente, Colin Reublinger, Andrew Robinson, Dan Schulte, Kumiko Sekioka, Catherine Shih, Howard Siegelman, Ivan Sorrentino, Ian Sutherland, Alcione Tavares, Koen Van Landeghem, Sergio Varela, and Ellen Zlotnick.

In addition, the authors would like to thank Colin Hayes and Jeremy Mynott for making the project possible in the first place. Most of all, very special thanks are due to Mary Vaughn for her dedication, support, and professionalism. Helen Sandiford would like to thank her family and especially her husband, Bryan Swan, for his support and love.

Welcome to Touchstone!

本系列教程的成功开发得益于剑桥国际语料库北美语料库丰富的资源和强大的功能。该大型语料库从日常对话、广播、电视节目、报纸及图书中广泛取材。

本系列教程利用计算机软件对该语料库进行分析，总结出英语的实际用法。我们以该语料库为标准，确保学生在每一课都能学到纯正、地道的英语。该语料库帮助我们遴选重点语法、词汇以及成功进行英语交流所必需的会话策略。

本系列教程会使你的英语学习妙趣横生。它为你提供与同学互动的多种机会。你们可以交换个人信息、进行班级问卷调查、角色扮演部分场景、做游戏，还可以讨论个人感兴趣的话题。使用本系列教程能让你逐步树立理解地道英语、在日常交流中清楚而有效地表达自我的信心。

希望大家能够喜欢本系列教程，并祝愿大家的英语课堂充满乐趣！

Michael McCarthy
Jeanne McCarten
Helen Sandiford

Unit features

Getting started presents new grammar in natural contexts such as surveys, interviews, conversations, and phone messages.

Figure it out challenges you to notice how grammar works.

Grammar is presented in clear charts.

Grammar exercises give you practice with new structures and opportunities to exchange personal information with your classmates.

Speaking naturally helps you understand and use natural pronunciation and intonation.

Building language builds on the grammar presented in Lesson A.

In conversation panels tell you about the grammar and vocabulary that are most frequent in spoken North American English.

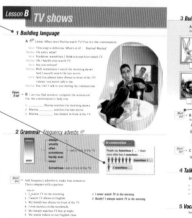

Building vocabulary uses pictures to introduce new words and expressions.

Word sort helps you organize vocabulary and then use it to interact with your classmates.

Talk about it encourages you to discuss interesting questions with your classmates.

Conversation strategy helps you "manage" conversations better. In this lesson, you learn how to ask questions that aren't too direct. The strategies are based on examples from the corpus.

Strategy plus teaches important expressions for conversation management, such as *I mean*, *Well*, and *Anyway*.

Listening and speaking skills are often practiced together. You listen to a variety of conversations based on real-life language. Tasks include "listen and react" activities.

Reading has interesting texts from newspapers, magazines, and the Internet. The activities help you develop reading skills.

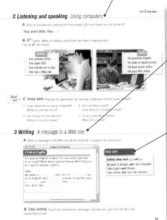

Writing tasks include e-mails, letters, short articles, and material for Web pages.

Help notes give you information on things like punctuation, linking ideas, and organizing information.

Vocabulary notebook is a page of fun activities to help you organize and write down vocabulary.

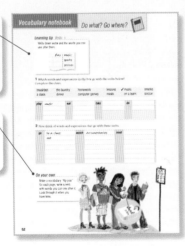

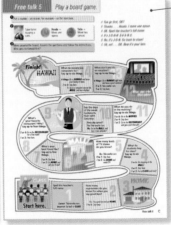

Free talk helps you engage in free conversation with your classmates.

On your own is a practical task to help you learn vocabulary outside of class.

Other features

A **Touchstone checkpoint** after every three units reviews grammar, vocabulary, and conversation strategies.

A **Self-study Audio CD/ CD-ROM** gives you more practice with listening, speaking, and vocabulary building.

The **Class Audio Program** presents the conversations and listening activities in natural, lively English.

The **Workbook** gives you language practice and extra reading and writing activities. **Progress checks** help you assess your progress.

Touchstone Level 1 Scope and sequence

	Functions / Topics	Grammar	Vocabulary	Conversation strategies	Pronunciation
Unit 1 **All about you** *pages 1–10*	• Say hello and good-bye • Introduce yourself • Exchange personal information (names, phone numbers, and e-mail addresses) • Spell names • Thank people	• The verb *be* with *I, you,* and *we* in statements, *yes-no* questions, and short answers • Questions with *What's . . . ?* and answers with *It's . . .*	• Expressions to say hello and good-bye • Numbers 0–10 • Personal information • Everyday expressions	• Ask *How about you?* • Use everyday expressions like *Yeah* and *Thanks*	• Letters and numbers • E-mail addresses
Unit 2 **In class** *pages 11–20*	• Ask and say where people are • Name personal items and classroom objects • Ask and say where things are in a room • Make requests • Give classroom instructions • Apologize	• The verb *be* with *he, she,* and *they* in statements, *yes-no* questions, and short answers • Articles *a, an,* and *the* • *This* and *these* • Noun plurals • Questions with *Where . . . ?* • Possessives *'s* and *s'*	• Personal items • Classroom objects • Prepositions and expressions of location	• Ask for help in class • Respond to *Thank you* and *I'm sorry*	• Noun plural endings
Unit 3 **Favorite people** *pages 21–30*	• Talk about favorite celebrities • Describe people's personalities • Talk about friends and family	• Possessive adjectives • The verb *be* in statements, *yes-no* questions, and short answers (summary) • Information questions with *be*	• Types of celebrities • Basic adjectives • Adjectives to describe personality • Family members • Numbers 10–101	• Show interest by repeating information and asking questions • Use *Really?* to show interest or surprise	• *Is he . . . ?* or *Is she . . . ?*

Touchstone checkpoint Units 1–3   pages 31–32

	Functions / Topics	Grammar	Vocabulary	Conversation strategies	Pronunciation
Unit 4 **Everyday life** *pages 33–42*	• Describe a typical morning in your home • Discuss weekly routines • Get to know someone • Talk about lifestyles	• Simple present statements, *yes-no* questions, and short answers	• Verbs for everyday activities • Days of the week • Time expressions for routines	• Say more than *yes* or *no* when you answer a question • Start answers with *Well* if you need time to think, or if the answer isn't a simple *yes* or *no*	• *-s* endings of verbs
Unit 5 **Free time** *pages 43–52*	• Discuss free-time activities • Talk about TV shows you like and don't like • Talk about TV-viewing habits	• Simple present information questions • Frequency adverbs	• Types of TV shows • Free-time activities • Time expressions for frequency • Expressions for likes and dislikes	• Ask questions in two ways to be clear and not too direct • Use *I mean* to repeat your ideas or to say more	• *Do you . . . ?*
Unit 6 **Neighborhoods** *pages 53–62*	• Describe a neighborhood • Ask for and tell the time • Make suggestions • Discuss advertising	• *There's* and *There are* • Quantifiers • Adjectives before nouns • Telling time • Suggestions with *Let's*	• Neighborhood places • Basic adjectives • Expressions for telling the time	• Use *Me too* or *Me neither* to show you have something in common with someone • Respond with *Right* or *I know* to agree with someone, or to show you are listening	• Word stress

Touchstone checkpoint Units 4–6   pages 63–64

Listening	Reading	Writing	Vocabulary notebook	Free talk
• Recognize responses to hello and good-bye *Memberships* • Listen for personal information, and complete application forms	• Different types of identification cards and documents	• Complete an application	*Meetings and greetings* • Write new expressions with their responses	*Meet a celebrity.* • Class activity: Introduce yourself and complete name cards for three "celebrities"
Who's absent? • Listen to a classroom conversation, and say where students are *Following instructions* • Recognize classroom instructions	• Classroom conversations	• Write questions about locations	*My things* • Link things with places	*What do you remember?* • Pair work: How much can you each remember about a picture?
Friends • Listen to three people's descriptions of their friends, and fill in the missing words	• A family tree	• Write questions about people	*All in the family* • Make a family tree	*Talk about your favorite people.* • Pair work: Score points for each thing you say about your favorite people

Touchstone checkpoint Units 1–3   pages 31–32

Listening	Reading	Writing	Vocabulary notebook	Free talk
What's the question? • Listen to answers and infer the questions *Teen habits* • Listen for information in a conversation, and complete a chart about a teenager's habits	*In the lifetime of an average American . . .* • A magazine article describing how much time people spend on daily activities over a lifetime	• Write an e-mail message about a classmate • Use capital letters and periods	*Verbs, verbs, verbs* • Draw and label simple pictures of new vocabulary	*Interesting facts* • Class survey: Ask questions to compare your classmates with the average New Yorker
What do they say next? • Listen to conversations and predict what people say next *Using computers* • Listen for the ways two people use their computers	*Are you an Internet addict?* • A magazine article and questionnaire about Internet use	• Write a message to a Web site about yourself • Link ideas with *and* and *but*	*Do what? Go where?* • Write verbs with the words you use after them	*Play a board game.* • Pair work: Do the activities and see who gets from class to Hawaii first
What's on this weekend? • Listen to a radio broadcast for the times and places of events *City living* • Listen for topics in a conversation, and then react to statements	*Classifieds* • A variety of classified ads from a local newspaper	• Write an ad for a bulletin board • Use prepositions for time and place: *between, through, at, on, for,* and *from . . . to . . .*	*A time and a place . . .* • Link times of the day with activities	*Find the differences.* • Pair work: List all the differences you find between two neighborhoods

Touchstone checkpoint Units 4–6   pages 63–64

	Functions / Topics	Grammar	Vocabulary	Conversation strategies	Pronunciation
Unit 7 **Out and about** pages 65–74	- Describe the weather - Leave phone messages - Talk about sports and exercise - Say how your week is going - Give exercise advice	- Present continuous statements, *yes-no* questions, short answers, and information questions - Imperatives	- Seasons - Weather - Sports and exercise with *play, do,* and *go* - Common responses to good and bad news	- Ask follow-up questions to keep a conversation going - React with expressions like *That's great!* and *That's too bad.*	- Stress and intonation in questions
Unit 8 **Shopping** pages 75–84	- Talk about clothes - Ask for and give prices - Shop for gifts - Discuss shopping habits	- *Like to, want to, need to,* and *have to* - Questions with *How much . . . ?* - *This, these; that, those*	- Clothing and accessories - Jewelry - Colors - Shopping expressions - Prices - "Time to think" expressions - "Conversation sounds"	- Take time to think using *Uh, Um, Well, Let's see,* and *Let me think* - Use "sounds" like *Uh-huh* to show you are listening, and *Oh* to show your feelings	- *Want to* and *have to*
Unit 9 **A wide world** pages 85–94	- Give sightseeing information - Talk about countries you want to travel to - Discuss international foods, places, and people	- *Can* and *can't*	- Sightseeing activities - Countries - Regions - Languages - Nationalities	- Explain words using *a kind of, kind of like,* and *like* - Use *like* to give examples	- *Can* and *can't*

Touchstone checkpoint Units 7–9   pages 95–96

	Functions / Topics	Grammar	Vocabulary	Conversation strategies	Pronunciation
Unit 10 **Busy lives** pages 97–106	- Ask for and give information about the recent past - Describe the past week - Talk about how you remember things	- Simple past statements, *yes-no* questions, and short answers	- Simple past irregular verbs - Time expressions for the past - Fixed expressions	- Respond with expressions like *Good luck, You poor thing,* etc. - Use *You did?* to show that you are interested or surprised, or that you are listening	- *-ed* endings
Unit 11 **Looking back** pages 107–116	- Describe experiences such as your first day of school or work - Talk about a vacation - Tell a funny story	- Simple past of *be* in statements, *yes-no* questions, and short answers - Simple past information questions	- Adjectives to describe feelings - Expressions with *go* and *get*	- Show interest by answering a question and then asking a similar one - Use *Anyway* to change the topic or end a conversation	- Stress and intonation in questions and answers
Unit 12 **Fabulous food** pages 117–126	- Talk about food likes and dislikes and eating habits - Make requests and offers - Invite someone to a meal - Make recommendations	- Countable and uncountable nouns - *How much . . . ?* and *How many . . . ?* - *Would you like (to) . . . ?* and *I'd like (to) . . .* - *Some* and *any* - *A lot of, much,* and *many*	- Foods and food groups - Expressions for eating habits - Adjectives to describe restaurants	- Use *or something* and *or anything* to make a general statement - End *yes-no* questions with *or . . . ?* to be less direct	- *Would you . . . ?*

Touchstone checkpoint Units 10–12   pages 127–128

Listening	Reading	Writing	Vocabulary notebook	Free talk
How's your week going? ▪ Listen to people talk about their week, and react appropriately *Do you enjoy it?* ▪ Listen to conversations and identify what type of exercise each person does and why he or she enjoys it	*Don't wait – just walk!* ▪ An article about the benefits of walking for exercise	▪ Write a short article giving advice about exercise ▪ Use imperatives to give advice	*Who's doing what?* ▪ Write new words in true sentences	*What's hot? What's not?* ▪ Group work: Discuss questions about current "hot" topics
I'll take it. ▪ Listen to conversations in a store, and write the prices of items and which items people buy *Favorite places to shop* ▪ Listen to someone talk about shopping, and identify shopping preferences and habits	*Shopping around the world* ▪ An article about famous shopping spots around the world	▪ Write a recommendation for a shopper's guide ▪ Link ideas with *because* to give reasons	*Nice outfit!* ▪ Label pictures with new vocabulary	*How do you like to dress?* ▪ Class activity: Survey classmates about the things they like to wear
National dishes ▪ Listen to a person talking about international foods, and identify the foods she likes *What language is it from?* ▪ Listen to a conversation, and identify the origin and meaning of words	*The travel guide* ▪ A page from a travel Web site with information, pictures, and travel advice	▪ Write a paragraph for a Web page for tourists ▪ Use commas in lists	*People and nations* ▪ Group new vocabulary in two ways	*Where in the world . . . ?* ▪ Pair work: Name different countries or cities where you can do interesting things

Touchstone checkpoint Units 7–9   pages 95–96

Listening	Reading	Writing	Vocabulary notebook	Free talk
What a week! ▪ Listen to people describe their week, and choose a response *Don't forget!* ▪ Listen for how people remember things, and identify the methods they use	*Ashley's journal* ▪ A week in Ashley's life from her personal journal	▪ Write a personal journal ▪ Order events with *before*, *after*, *when*, and *then*	*Ways with verbs* ▪ Write down information about new verbs	*Yesterday . . .* ▪ Pair work: Use the clues in a picture to "remember" what you did yesterday
Weekend fun ▪ Listen to a conversation about last weekend, and identify main topics and details *Funny stories* ▪ Listen to two stories, identify the details, and then predict the endings	*Letters from our readers* ▪ A letter telling a funny story about a reader's true experience	▪ Complete a funny story ▪ Use punctuation to show direct quotations or speech	*Past experiences* ▪ Use a time chart to log new vocabulary	*Guess where I went on vacation.* ▪ Group work: Ask and answer questions to guess where each person went on vacation
Lunchtime ▪ Listen to people talking about lunch, and identify what they want; then react to statements *Do you recommend it?* ▪ Listen to someone tell a friend about a restaurant, and identify important details about it	*Restaurant guide* ▪ Restaurant descriptions and recommendations	▪ Write a restaurant review ▪ Use adjectives to describe restaurants	*I love to eat!* ▪ Group vocabulary by things you like and don't like	*Do you live to eat or eat to live?* ▪ Class activity: Survey classmates to find out about their eating habits

Touchstone checkpoint Units 10–12   pages 127–128

Useful language for . . .

Getting help

What's the word for "_____" in English?

How do you spell "_____"?

What does "_____" mean?

I'm sorry. Can you repeat that, please?

Can you say that again, please?

Can you explain the activity again, please?

Working with a partner

I'm ready. Are you ready?

No. Just a minute.

You go first.

OK. I'll go first.

What do you have for number 1?

I have . . .

Do you want to be A or B?

I'll be A. You can be B.

Let's do the activity again.

OK. Let's change roles.

That's it. We're finished.

What do we do next?

Can I read your paragraph?

Sure. Here you go.

All about you

In Unit 1, you learn how to . . .

- ■ use the verb *be* with *I*, *you*, *we*, and *it*.
- ■ say hello and good-bye.
- ■ say your name, telephone number, and e-mail address.
- ■ ask *How about you?*
- ■ use everyday expressions like *Thanks*.

Before you begin . . .

Match each expression with a picture.

| 1 Hello. | ☐ Thanks. | ☐ Good morning. | ☐ Hi. |
| ☐ Bye. | ☐ Good night. | ☐ Thank you. | ☐ Good-bye. |

Hello and good-bye

Matt   *Good morning, Sarah. How are you?*
Sarah *Good. How are you, Matt?*
Matt   *I'm fine, thanks.*

Matt   *Hello. I'm Matt Lenski.*
Emily *Hi, I'm Emily Kim. Nice to meet you.*
Matt   *Nice to meet you.*

1 Getting started

A Listen. Matt and Sarah are friends. Are Matt and Emily friends?
Practice the conversations.

Figure it out

B Can you complete these conversations? Then practice with a partner.
Use your own names.

1 A Hello. I **'m** Chris.
   B Hi. Nice to meet you. _____ Sam.
   A Nice to meet _____ .

2 A Hi, Pat. How _____ you?
   B I'm _____ . How are _____ ?
   A Good, thanks.

2 Building vocabulary

A Listen. Practice the conversations.

> **Emily** *Good night.*
> **Matt** *Good night. Have a good evening.*
> **Emily** *Thank you. You too.*

> **Sarah** *Bye. See you tomorrow.*
> **Matt** *Bye. See you.*

B Listen to the conversations. Check (✓) the responses you hear.

1. Bye. Have a good evening.
   - ☐ You too. Good-bye.
   - ✓ You too. Good night.

2. Hi. How are you?
   - ☐ Good, thanks.
   - ☐ I'm fine.

3. Bye. See you later.
   - ☐ OK. See you later.
   - ☐ Bye. See you next week.

4. Hey, Oscar!
   - ☐ Hi. How are you?
   - ☐ Hello.

5. Good-bye. Have a nice day.
   - ☐ Thank you.
   - ☐ Thanks. You too.

Word sort → **C** Write three expressions for saying hello and good-bye. Compare with a partner.

How are you?

Hello.

Good-bye.

D *Class activity* Say hello and good-bye to five classmates.

3 Vocabulary notebook *Meetings and greetings*

See page 10 for a new way to log and learn vocabulary.

Hi. My name is David.
My **last name** is Hanson.

Name: *David Allen Hanson*
FIRST   MIDDLE   LAST
☑ single ☐ married

Hi, I'm Liz Park.
My **first name** is Elizabeth.
Liz is short for Elizabeth.

Name: *Elizabeth — Park*
FIRST   MIDDLE   LAST
☑ single ☐ married

I'm Mary Gomez.
My **middle name** is Ann.
Frank is my husband.

Name: *Mary Ann Gomez*
FIRST   MIDDLE   LAST
☐ single ☑ married

1 Saying names in English

A Listen to the people above give their names.

B Complete the sentences. Then compare with a partner.

1. My first name is _____ .
2. My last name is _____ .
3. My middle name is _____ .
4. My nickname is _____ .
5. My teacher's name is _____ .
6. My favorite name is _____ .

Miss, Mrs., Ms., Mr.?

- David Hanson is single.   → **Mr.** Hanson
- Liz Park is single.   → **Ms.** Park / **Miss** Park
- Mary Gomez is married.   → **Ms.** Gomez / **Mrs.** Gomez
- Frank Gomez is married.   → **Mr.** Gomez

C Listen and say the alphabet. Circle the letters in your first name.

A a	B b	C c	D d	E e	F f	G g	H h	I i	J j	K k	L l	M m
N n	O o	P p	Q q	R r	S s	T t	U u	V v	W w	X x	Y y	Z z

D Listen. Then practice the conversation with a partner. Use your own names.

A What's your name?
B Catherine Ravelli.
A How do you spell *Catherine*?
B C-A-T-H-E-R-I-N-E.
A Thanks. And your last name?
B R-A-V-E-L-L-I.

Katherine?
Kathryn?
Catherine?

About you → **E Class activity** Ask your classmates their names. Make a list.

4

2 Building language

A Listen. Which classroom is Carmen in this term? What about Jenny? Practice the conversation.

Mr. Martin	Good morning. Are you here for an English class?
Carmen	Yes, I am. I'm Carmen Rivera.
Mr. Martin	OK. You're in Room B.
Jenny	And I'm Jenny.
Mr. Martin	Are you Jenny Loo?
Jenny	No, I'm not. I'm Jenny Lim. Am I in Room B, too?
Mr. Martin	Yes. . . . Wait – no, you're not. You're in Room G.
Jenny	Oh, no! Carmen, we're not in the same class!

> **Figure it out** →

B Complete the answers. Then check the names of five classmates.

1 *A* Are you Amy?
  B Yes, I _____ .

2 *A* Are you Amy?
  B No, _____ not.

3 Grammar *The verb be: I, you, and we*

I'm Jenny.	**Are you** Jenny?
I'm not Carmen.	Yes, **I am**. / No, **I'm not**.
You're in Room G.	**Am I** in Room B?
You're not in Room B.	Yes, **you are**. / No, **you're not**.
We're in different classes.	**Are we** in the same class?
We're not in the same class.	Yes, **we are**. / No, **we're not**.
I'm = I am you're = you are we're = we are	

In conversation . . .

I is the most common word.
I'm is more common than *I am*.

I'm

I am

A Complete the conversations. Then practice with a partner.

1 *A* __Are__ you Emiko?
  B Yes, I _____ . I_____ here for an English class. _____ you here for English, too?
  A No, I____ _____ . I____ here for a French class.

2 *A* _____ you Chris?
  B Yes, I _____ . _____ we in the same class?
  A Yes, we _____ . I____ Dino.
  B Hi, Dino. Nice to meet you.

B *Pair work* Choose a conversation and practice. Use your own information. Then act out your conversation for the class.

1 Numbers 0–10

A 🎧 Listen and say the numbers.

0	1	2	3	4	5	6	7	8	9	10
zero	one	two	three	four	five	six	seven	eight	nine	ten

B 🎧 Listen. Then practice.

❶ *My passport number is 649-321-508.*

❷ *My ID number is 259-62-1883.*

❸ *My phone number is 216-555-7708. My e-mail address is dsmith6@cup.org.*

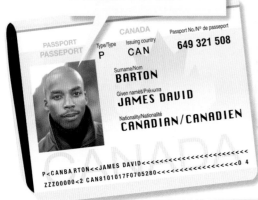

Numbers and e-mail addresses

216-555-7708 = "two-one-six, five-five-five, seven-seven-**oh (zero)**-eight"
dsmith6@cup.org = "d-smith-six-**at**-c-u-p-**dot**-org"

2 Building language

A 🎧 Listen. What is Victor's telephone number? Practice the conversation.

Receptionist	Hi! Are you a member?
Victor	No, I'm just here for the day.
Receptionist	OK. So, what's your name, please?
Victor	Victor Lopez.
Receptionist	And what's your phone number?
Victor	It's 646-555-3048.
Receptionist	And your e-mail address?
Victor	Um . . . it's vlopez6@cup.org.
Receptionist	OK. So it's $10 for today. Here's your pass.
Victor	Thanks.

Welcome to Fitness Gym

DAY MEMBERS WELCOME $10.00

Figure it out

B Can you complete these questions and answers? Then practice with a partner.

❶ *A* _____ your name?
B Joe Garrett.

❷ *A* What's _____ ?
B It's jgarrett@cup.org.

❸ *A* _____ ?
B _____ 646-555-4628.

3 *Grammar* *What's . . . ?; It's . . .*

What's your name?   **My name's** Victor Lopez.
What's your e-mail address?   **It's** vlopez6@cup.org.
What's your phone number?   **It's** 646-555-3048.

What's = What is   *name's = name is*   *It's = It is*

In conversation . . .

Phone is 6 times more common than **telephone**.

phone

telephone

A Match the questions and answers. Then practice.

1. What's your teacher's name? __b__
2. What's your first name? ____
3. What's your e-mail address? ____
4. What's your phone number? ____
5. What's your last name? ____

a. Rachel.
b. It's Ms. Gardino.
c. My last name? Yoshida.
d. It's yoyo3@cup.org.
e. 646-555-3907.

About you

B *Pair work* Ask and answer three questions with *What's*. Give your own answers.

"What's your teacher's name?"   *"It's Mr. Williams."*

4 *Listening and speaking* *Memberships*

A Listen to the conversations. Complete the application forms.

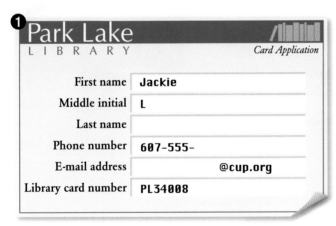

❶ **Park Lake**
L I B R A R Y   *Card Application*

First name	Jackie
Middle initial	L
Last name	
Phone number	607-555-
E-mail address	@cup.org
Library card number	PL34008

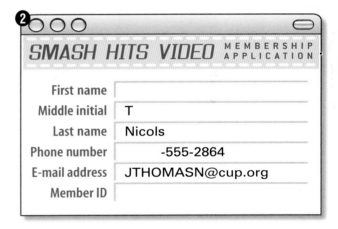

❷ **SMASH HITS VIDEO** MEMBERSHIP APPLICATION

First name	
Middle initial	T
Last name	Nicols
Phone number	-555-2864
E-mail address	JTHOMASN@cup.org
Member ID	

About you

B *Pair work* Now complete this form for a partner. Ask questions.

ENGLISH CLUB APPLICATION

First name	
Middle initial	
Last name	
Phone number	
E-mail address	

A **What's your first name?**
B **Silvia.**
A **How do you spell it? . . .**

Are you here for the concert?

Conversation strategy *How about you?*

A Can you complete the conversation with the questions in the box?

A _____ ?

B *Yes, I am.* _____ ?

A *Yes, me too.*

How about you?
Are you a new student?

Now listen. Are Alicia and Adam friends?

Alicia **It's a beautiful day.**

Adam **Yeah, it is.**

Alicia **Are you here for the concert?**

Adam **Yeah, I am. How about you?**

Alicia **Yeah, me too. So, are you a student here?**

Adam **Yeah. How about you?**

Alicia **No, I'm here on vacation.**

Adam **Nice. By the way, I'm Adam.**

Alicia **Hi, Adam. I'm Alicia.**

Notice how Adam uses *How about you?* to ask the same questions as Alicia.

"Are you here for the concert?"

"Yeah, I am. How about you?"

B Complete the conversations below. Then practice with a partner.

1 *Lora* Are you new here?

Ying Yes, I am. _____ ?

Lora Yes, me too.

2 *Marie* Hello. Are you here on vacation?

Koji Yes, I am. _____ ?

Marie No, I'm here on business.

SELF-STUDY
AUDIO CD
CD-ROM

8

2 *Strategy plus* *Everyday expressions*

Some everyday expressions are more formal.

How are you?

More formal	Less formal
Yes.	Yeah.
Thank you.	Thanks.
Hello.	Hi.
How are you?	How are you doing?
I'm fine.	OK. / Pretty good. / Good.
Good-bye.	Bye. / See you. / See you later.

How are you doing?

In conversation . . .

Yeah is 10 times more common than **yes**.

Yeah.

Yes.

A Complete these conversations with expressions from the box above. Compare with a partner.

➊

Kathy	Hi, sorry I'm late.
	_____ ?
Mike	Pretty good. How are you?
Kathy	_____ .

Later . . .

Kathy	Bye. See you.
Mike	_____ .

➋

Jeff	Good morning, Mrs. Swan.
	_____ ?
Mrs. Swan	_____ . How are you?
Jeff	I'm fine, _____ .

Later . . .

Jeff	Good-bye, Mrs. Swan.
Mrs. Swan	_____ .

B *Pair work* Practice the conversations.

3 *Free talk* *Meet a celebrity.*

See *Free talk 1* at the back of the book for more speaking practice.

Learning tip *Learning expressions*

Write new expressions with their responses, like this:

See you later.   Bye. See you.

Write a response for each expression.

1. Hello.

2. Good morning.

3. Hi. I'm Helen.

4. How are you?

5. Have a nice day.

6. See you tomorrow.

7. Have a good evening.

8. Good night.

On your own

Before your next class, say hello and good-bye (in English!) to three people.

Hi or Hello?

People say *Hi* and *Bye* more than *Hello* and *Good-bye*.

Hi.
Hello.
Bye.
Good-bye.

Hi. How are you?

In class

In Unit 2, you learn how to . . .

- use the verb *be* with *he*, *she*, *they*, *this*, and *these*.
- talk about things and places in a classroom.
- ask for help in class.
- respond to *Thank you* and *I'm sorry*.

Before you begin . . .

Where are these people?

Match the pictures with the sentences.

	He's at home.		They're in class.
1	She's at work.		They're at the library.

Miss Cass Where's Jun? Is he here today?
Ana No, he's not. Maybe he's at work.
Miss Cass OK. How about Laura?
Ana I don't know. I think she's sick.
Miss Cass Oh. OK. Are Kim and Phong here?
Ana No, they're in the cafeteria.
Miss Cass They're late again. OK. And Alan?
Ana He's over there. I think he's asleep!

Jun

Laura

Kim and Phong

1 Getting started

A Listen. Where are Ana's classmates today? Practice the conversation.

Figure it out → **B** Can you complete the questions and answers? Use the conversation above to help you.

❶ A Is Jun in class today?
   B No, _____ not.

❷ A Is Laura here today?
   B No, _____ sick.

❸ A _____ Kim and Phong in class?
   B No, _____ late.

2 *Grammar* *The verb be: he, she, and they*

Jun is at work.	**Jun is not** here.	**Is Jun** here?
He's at work.	**He's not** here.	Yes, **he is**. / No, **he's not**.
Laura's sick.	**Laura's not** in class.	**Is she** sick?
She's sick.	**She's not** in class.	Yes, **she is**. / No, **she's not**.
Kim and Phong are late.	**Kim and Phong are not** here.	**Are they** late?
They're late.	**They're not** here.	Yes, **they are**. / No, **they're not**.

Laura's = Laura is   *He's = He is*
  She's = She is   *They're = They are*

▶ **In conversation . . .**

People usually shorten *is* to **'s** after names.
Jun's at work. Laura's not in class.

A These people are also Ana's classmates. Where are they today? Complete the sentences.

 ❶

 ❷

 ❸

David _____
_____ .

Connie and Dan _____
_____ .

Sue and Min Ji _____
_____ .

B Complete the questions. Then ask and answer the questions with a partner.

1. __Is__ David sick?
2. _____ Dan in class?
3. _____ Connie and Dan at home?
4. _____ Min Ji at work?
5. _____ Sue at the library?
6. _____ Sue and Min Ji in class?

"Is David sick?"   *"No, he's not. He's in class."*

3 *Listening* *Who's absent?*

A Listen. It's the next day. Where are these students today? Match each student with a place.

1. Jun's __c__
2. Kim's _____
3. Laura's _____
4. David's _____

a. at the library.
b. at work.
c. in the cafeteria.
d. at home.

 About you ➡

B *Pair work* Ask and answer questions about your classmates.

"Is Lisa sick today?"   *"Yes, she is. She's at home."*

13

1 Building vocabulary

A 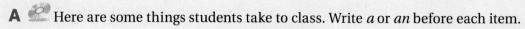 Here are some things students take to class. Write *a* or *an* before each item.
Then listen and say the words. Check your answers.

Articles

a + consonant sound
   a **b**ag
an + vowel sound
   *an e*raser

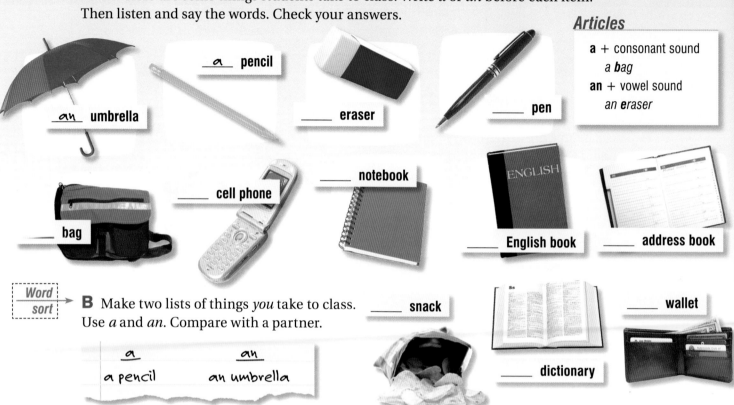

a pencil

_____ eraser

_____ pen

an umbrella

_____ cell phone

_____ notebook

_____ English book

_____ address book

_____ bag

_____ snack

_____ dictionary

_____ wallet

Word sort **B** Make two lists of things *you* take to class.
Use *a* and *an*. Compare with a partner.

a	*an*
a pencil	an umbrella

2 Building language

A    Listen. Which things are Bill's? Practice the conversations.

Andy What's this?
Bill It's an MP3 player.
   It's my new "toy."

Michi Is this your watch, Bill?
Bill Oh, yes, it is. Thanks.
Michi And are these your glasses?
Bill Yes, they are!

Scott Excuse me. Are these
   your keys?
Bill Um . . . no, they're not.
   These are my keys right here.

Figure it out **B** Can you complete these questions? Use your own ideas.
Ask and answer your questions with a partner.

❶ Is this your _____ ?   ❷ Are these your _____ ?

3 Grammar *This and these; noun plurals*

		Regular plurals		**Irregular plurals**	
This is an MP3 player.	**These are** sunglasses.	bag	bag**s**	man	men
What's this?	**What are these?**	watch	watch**es**	woman	wom**e**n
It's an MP3 player.	**They're** sunglasses.	dictionary	dictionar**ies**	child	child**ren**
		key	key**s**		
Is this your watch?	**Are these** your keys?				
Yes, **it is**.	Yes, **they are**.	**Some nouns are only plural:**			
No, **it's not**.	No, **they're not**.	*jeans, scissors, glasses, sunglasses*			

Complete the questions and answers about the pictures. Then practice with a partner.

❶

A What**'s this** ?
B **I think it's a cell phone** .

❷

A Is _____ your _____ ?
B No, _____ .

❸

A What _____ ?
B _____ .

❹

A Are _____ your _____ ?
B Yes, _____ .

❺

A Are _____ your _____ ?
B No, _____ .

❻

A What _____ ?
B _____ .

4 Speaking naturally *Noun plural endings*

/s/ wallet**s**, book**s**	/z/ pen**s**, key**s**	/ɪz/ watch**es**, orang**es**

A Listen and repeat the words above. Notice the noun plural endings.

B Listen. Do the nouns end in /s/, /z/, or /ɪz/? Check (✓) the correct column.

What's in your bag?	/s/	/z/	/ɪz/
1. *three textbooks*	✓	☐	☐
2. *two cell phones*	☐	☐	☐
3. *four snacks*	☐	☐	☐
4. *my sunglasses*	☐	☐	☐
5. *five credit cards*	☐	☐	☐

About you → **C Group work** Tell the group what's in your bag. Who has something unusual?

"What's in your bag, Carlos?"   *"A wallet, two keys, . . ."*

15

In the classroom

- [ ] a map
- [ ] some posters
- [ ] a CD player
- [ ] a table
- [ ] some chairs
- [ ] a computer
- [ ] a desk
- [ ] a clock
- [ ] a board
- [ ] some videos
- [ ] a wastebasket
- [ ] a calendar
- [ ] a TV
- [ ] a VCR
- [ ] some dictionaries

1 Building vocabulary

A Listen and say the words above. Which things are in your classroom? Check (✓) the boxes. What else is in your classroom?

Word sort →

B Look around your classroom. What things are in these places? Write the words below the pictures.

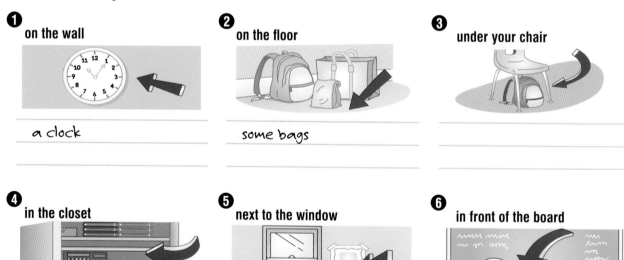

❶ **on the wall**

a clock

❷ **on the floor**

some bags

❸ **under your chair**

❹ **in the closet**

❺ **next to the window**

❻ **in front of the board**

C *Pair work* Ask and answer questions about your classroom.

"What's on the wall?"   *"A clock, a map, and some posters . . ."*

2 Building language

A Listen. What is the teacher looking for? Practice the conversation.

Mrs. Evans OK, so . . . where's the VCR?
    Paula It's in the closet.
Mrs. Evans Oh, right. And the videos? Where are they?
    Paula They're on the desk, under your coat.
Mrs. Evans OK, um . . . and where are the students'
              homework papers?
    Paula They're on the floor.
Mrs. Evans Uh-oh, what's this under my foot?
    Paula It's Mario's homework.
Mrs. Evans Oops! . . . Uh, where are my glasses?
              They're not on my desk.
    Paula Uh . . . they're on your head!

Figure it out

B Can you complete the questions? Then ask and answer the questions with a partner.

1. Where _____ the VCR?
2. Where _____ the teacher's coat?
3. Where _____ the students' papers?

3 Grammar *Questions with Where; possessives 's and s'*

Where's Mario's homework?	It's on the floor.	Mario**'s** homework
Where's the teacher's coat?	It's on the desk.	the teacher**'s** glasses
Where are the students' papers?	They're on the floor.	three student**s'** papers
Where's = Where is		

A *Pair work* Ask and answer these questions about the classroom on page 16.
Can you ask four more questions?

1. Where's the teacher's desk?
2. Where's the TV?
3. Where's the teacher's chair?
4. Where are the students' dictionaries?
5. Where's the computer?
6. Where are the posters?

"Where's the teacher's desk?"   *"It's in front of the board."*

About you

B Write four questions about things in your classroom. Use these ideas or add your own. Then ask a partner your questions.

the teacher's books   **the teacher's bag**   **the students' bags**   **the wastebasket**

4 Vocabulary notebook *My things*

See page 20 for a new way to log and learn vocabulary.

What's the word for this in English?

1 Conversation strategy *Asking for help in class*

A Can you match the questions and answers?

1. How do you spell book? _____
2. Can I borrow your pen? _____
3. What's the word for this in English? _____

a. Sure.
b. Pencil.
c. B-O-O-K.

Now listen. How many questions does Ming-wei ask?

Ming-wei	**Excuse me, what's the word for this in English?**
Sonia	**Highlighter.**
Ming-wei	**Thanks.**
Sonia	**Sure.**
Ming-wei	**Uh . . . how do you spell it?**
Sonia	**I don't know. Sorry.**
Ming-wei	**That's OK. Thanks anyway. . . . Can I borrow a pen, please?**
Sonia	**Sure. Here you go.**
Ming-wei	**Thank you.**
Sonia	**You're welcome.**
Ms. Larsen	**OK. Open your books to page 4.**
Ming-wei	**Excuse me, can you repeat that, please? What page?**
Ms. Larsen	**Sure. Page 4.**

Notice how Ming-wei asks for help in class. Find his questions.

"What's the word for this in English?"
"How do you spell it?"

B Complete the conversations. Then practice with a partner.

1
A What's this?
B It's an eraser.
A Oh, _____ ?
B E-R-A-S-E-R.

2
A Excuse me, can I _____ your dictionary, please?
B Sure. Here you go.

3
A What's your phone number?
B _____ , please?
A Yes. What's your phone number?

About you → **C** *Pair work* Practice the conversations again. Use your own ideas.

SELF-STUDY
AUDIO CD
CD-ROM

2 *Strategy plus* *Common expressions and responses*

Here are some responses to
Thank you **and** *I'm sorry*:

When people say . . .	You can say . . .
Thank you.	*You're welcome.*
Thanks.	*Sure.*
I'm sorry.	*That's OK.*
I'm sorry. I don't know.	*That's OK. Thanks anyway.*

`I'm sorry.`

`That's OK.`

Circle the correct response. Then practice with a partner.

1 *A* Can I borrow your pen, please?
   B **Sure.** **/ I don't know.**
   A Thanks.
   B **Thanks anyway. / You're welcome.**

2 *A* You're late.
   B **I'm sorry. / Thanks.**
   A That's OK.

3 *A* What's the word for this?
   B I don't know. Sorry.
   A **Sure. / That's OK.** What about this?
   B I don't know.
   A OK. **Thanks anyway. / You're welcome.**

3 *Listening and speaking* *Following instructions*

A Match the pictures with the instructions. Then listen to the conversations, and check your answers.

a. *Listen to the conversation.*

b. *Answer the questions on page 9.*

c. *Turn to page 7, and look at Exercise 1.*

d. *Write the word **eraser** in your notebook.*

B *Pair work* Give and follow four instructions. Ask for help if you need it.

A Look at the picture on page 8.
B Can you repeat that, please?
A Sure. Look at the picture on page 8.

4 *Free talk* *What do you remember?*

See *Free talk 2* at the back of the book for more speaking practice.

My things

Learning tip *Linking things with places*

Make lists of things you keep in different places.

> in my bag – my wallet, my keys

1 Label the things on the desk.

some books

2 Now make lists of your things.

What's in your bag?

What's in your wallet?

What's under your desk?

What's in your pockets?

On your own

Find a magazine with pictures of things. Label the pictures. How many words can you label?

TV
window
sofa

3

Favorite people

In Unit 3, you learn how to . . .

- use *my*, *your*, *his*, *her*, *our*, and *their*.
- use the verb *be* in information questions.
- talk about your favorite celebrities, friends, and family.
- show interest in a conversation.
- use *Really?* to show interest or surprise.

Before you begin . . .

Match the pictures with the sentences.

	He's a singer.		They're soccer players.
	She's an actor.	1	He's an artist.

For each sentence, think of someone you know.

Celebrities

Sean Penn

Norah Jones

the Williams sisters

Sandra I love these shows about celebrities. Hmm. Who's that guy? Oh, look. It's Sean Penn. He's so good-looking. His new movie is great.

. . . Oh, and there's Norah Jones. She's my favorite singer. Her voice is amazing.

. . . And look – the Williams sisters, my favorite tennis players. Their matches are always exciting. You're a tennis fan, right? John? . . . John? Wake up!

1 Getting started

A Listen. Sandra is watching TV with John. Is the show interesting for John?

Figure it out

B Can you complete the sentences? Use the information above to help you.

1. Sean Penn is an actor. _____ movies are very good.
2. Norah Jones is a famous singer. _____ new video is great.
3. The Williams sisters are tennis players. _____ matches are always great.

2 Grammar *Be in statements; possessive adjectives*

I'm	a Sean Penn fan.	**My**	favorite actor is Sean Penn.
You're	a tennis fan.	**Your**	favorite sport is tennis.
He's	an actor.	**His**	new movie is great.
She's	a famous singer.	**Her**	voice is amazing.
We're	Giants fans.	**Our**	favorite team is the Giants.
They're	tennis players.	**Their**	matches are exciting.

Circle the correct words to complete the conversations.
Then practice with a partner.

1. *A* (**I'm**)/ **My** a Sting fan.
   B Yeah, **he's / his** music is amazing.
   A You know, **he's / his** real name is
   Gordon Matthew Sumner.

2. *A* **I'm / My** favorite band is Black Eyed Peas.
   B Oh, **they're / their** very good.
   A You know, **they're / their** new CD is out now.

3. *A* Nicole Kidman's new movie is really great.
   B Yeah? **She's / Her** movies are always good.
   A I know. **She's / Her** my favorite actor.

4. *A* What's **you're / your** favorite show?
   B **I'm / My** favorite show? *Friends*.
   A Yeah. It's **we're / our** favorite show, too.
   In our family, **we're / our** all *Friends* fans.

3 Talk about it *My favorite celebrities*

Write the names of your favorite celebrities below. Then talk about them with
a partner. How many things can you say?

actor	Johnny Depp	band	
singer		team	
writer		artist	

"My favorite actor is Johnny Depp. He's so good-looking. His new movie is great."

People we know

She's **smart**. She's very **interesting**.

He's **quiet** and **shy**.

She's **friendly** and **outgoing**.

They're very **nice**. They're **fun**.

He's **lazy**.

1 Building vocabulary

A Listen and say the sentences. Do you have friends like these? Tell the class.

Word sort

B How many words can you think of to describe people you know? Complete the chart. Then tell a partner.

my friends	my best friend	my neighbor
very smart		

"My friends are very smart. They're . . ."

2 Building language

A Listen. What is Tim's new boss like? Practice the conversation.

Dana So, how's your new job? Are you busy?

Tim Yes. It's hard work, you know. I'm tired.

Dana Really? What are your co-workers like? Are they nice?

Tim Yes, they are. They're really friendly.

Dana Great. And is your boss OK?

Tim She is, yeah. She's nice. Um . . . she's not very strict.

Dana Good, because you're late for work.

Figure it out

B Can you complete these questions and answers? Use your own ideas. Then compare with a partner.

❶ *A* How's your new teacher?
_____ she _____ ?
B Yes, she _____ .

❷ *A* What about your classmates?
_____ they _____ ?
B Yes, they _____ .

❸ *A* And what's class like?
_____ it _____ ?
B No, it's not.

3 Grammar Yes-No questions and answers; negatives

Am I late?	Yes, **you are.**	No, **you're not.**	**You're not** late.
Are you busy?	Yes, **I am.**	No, **I'm not.**	**I'm not** busy.
Is he tired?	Yes, **he is.**	No, **he's not.**	**He's not** tired.
Is she strict?	Yes, **she is.**	No, **she's not.**	**She's not** strict. (My boss **isn't** strict.)
Is it hard work?	Yes, **it is.**	No, **it's not.**	**It's not** hard work.
Are we late?	Yes, **we are.**	No, **we're not.**	**We're not** late.
Are they nice?	Yes, **they are.**	No, **they're not.**	**They're not** nice. (My co-workers **aren't** nice.)

About you → Write *yes-no* questions. Then write true answers.
Ask and answer the questions with a partner.

> **In conversation . . .**
>
> People use **'s not** and **'re not** after pronouns.
> She**'s not** strict.
> They**'re not** nice.
>
> **Isn't** and **aren't** often follow nouns.
> My boss **isn't** strict.
> My co-workers **aren't** nice.

1. you / shy ?
   __Are you shy?__   __Yes, I am.__

2. this class / easy ?
   _____   _____

3. the teacher / strict ?
   _____   _____

4. the students in this class / lazy ?
   _____   _____

5. your neighbors / nice ?
   _____   _____

6. your friends / outgoing ?
   _____   _____

4 Speaking naturally Is he…? or Is she…?

/ɪziy/
Is he a student?

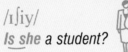
/ɪʃiy/
Is she a student?

A Listen and repeat the questions above. Notice the pronunciation of *Is he . . . ?* and *Is she . . . ?*

B Listen. Do you hear *Is he . . . ?* or *Is she . . . ?* Circle *he* or *she.*

1. Is **he / she** a friend from high school?
2. Is **he / she** a college student?
3. Is **he / she** shy?
4. Is **he / she** smart?
5. Is **he / she** interesting?
6. Is **he / she** fun?

About you → **C** *Pair work* Find out about your partner's best friend.
Ask and answer questions like the ones above.

A **Is he a friend from high school?**
B **No, he's a neighbor.**

Interesting? Smart? Shy?

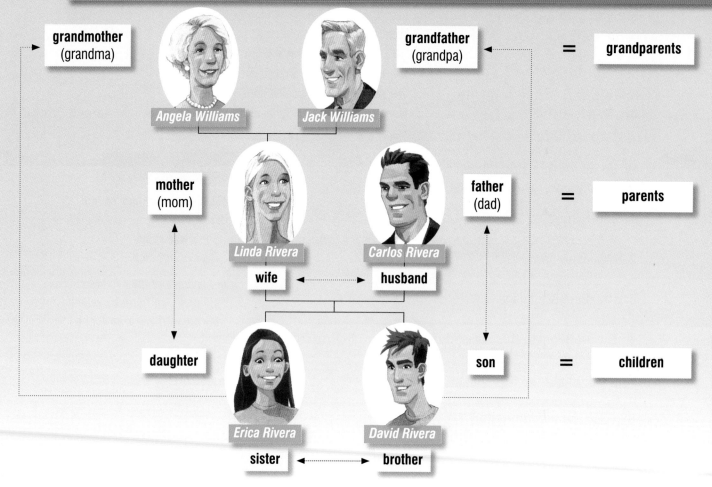

grandmother (grandma)   grandfather (grandpa)   =   grandparents

Angela Williams   Jack Williams

mother (mom)   father (dad)   =   parents

Linda Rivera   Carlos Rivera

wife ↔ husband

daughter   son   =   children

Erica Rivera   David Rivera

sister ↔ brother

1 Building vocabulary

A 🔊 Listen and say the words above. Then with a partner, ask and answer questions about the people. How many answers can you think of for each person?

"Who's Angela?"   "She's Jack's wife. She's Linda's mother. She's David's grandmother."

B 🔊 Listen and say the numbers. Do you have any "lucky numbers"? Tell the class.

10 ten	**16** sixteen	**22** twenty-two	**28** twenty-eight	**70** seventy
11 eleven	**17** seventeen	**23** twenty-three	**29** twenty-nine	**80** eighty
12 twelve	**18** eighteen	**24** twenty-four	**30** thirty	**90** ninety
13 thirteen	**19** nineteen	**25** twenty-five	**40** forty	**100** one hundred
14 fourteen	**20** twenty	**26** twenty-six	**50** fifty	**101** a hundred and one
15 fifteen	**21** twenty-one	**27** twenty-seven	**60** sixty	

About you → **C** *Pair work* Student A: Tell your partner the names and ages of your family members. Student B: Write the information you hear. Then check the information with your partner.

A **My mother's name is Sandra. She's fifty-five.**

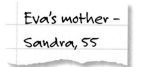
Eva's mother – Sandra, 55

➡ *B* **Eva, is your mother's name Sandra?** *A* **Yes, it is.**

26

2 *Building language*

A Listen. How old are Erica's grandparents? Practice the conversation.

Akemi So, who's this?

Erica My grandma. And this is my grandpa. He's a nice man. He's seventy-eight now.

Akemi Really? And how old is your grandmother?

Erica She's seventy-two.

Akemi She's very pretty. What's her name?

Erica Angela.

Akemi That's a nice name. So, where are your grandparents from originally?

Erica They're from Texas.

Figure it out → **B** Can you put the words in the correct order to make questions? Then ask and answer the questions with a partner.

1. are / from / parents / originally / where / your ?
2. your / is / old / father / how ?
3. names / your / what / grandparents' / are ?

3 *Grammar* *Information questions with be*

How are you?	Who's this?	How are your parents?
I'm fine.	It's my grandmother.	They're fine, thanks.
Where are you from?	**Where's she from?**	**Where are they today?**
I'm from Florida.	She's from Texas.	They're at home.
How old are you?	**What's she like?**	**What are their names?**
Twenty-three.	She's very smart.	Linda and Carlos.

A Write at least six questions to ask your classmates about their families.

What . . . ?	Where . . . ?	How . . . ?
What's your father like?		

About you → **B** *Class activity* Ask three classmates your questions.

"What's your father like?"   *"He's very outgoing."*

4 *Vocabulary notebook* *All in the family*

See page 30 for a new way to log and learn vocabulary.

SELF-STUDY
AUDIO CD
CD-ROM

1 Conversation strategy *Showing interest*

A Can you add a question to show you're interested in the conversation?

    A *Here's a picture of my best friend.*

    B *Really? _____ ?*

    Now listen. What do you find out about Eve's friend?

Mark	This is a great photo. Who is it?
Eve	It's a friend of mine – Natasha.
Mark	Oh? Where's she from?
Eve	She's from London, but she's here in Miami now.
Mark	London? Wow. Is she a student here?
Eve	No, she's an artist – a painter. She's an amazing woman.
Mark	A painter? Really? What are her paintings like?
Eve	They're wonderful. Look.
Mark	Oh. Interesting. . . . Um, what is it?

Notice how Mark shows interest. He repeats words and asks questions. Find examples in the conversation.

"She's from London, but she's here in Miami now."
"London? Wow. Is she a student here?"

B Complete the responses. Then practice with a partner.

1   A   My friend Gemma is a singer.
      B   <u> A singer </u> ? Is she in a band?

2   A   My best friend's name is Vlad.
      B   _____ ? Where's he from?

3   A   My friends Joshua and Pat are actors.
      B   _____ ? Are they famous?

C *Pair work* Student A: Tell your partner about a friend.
Student B: Ask questions to show interest. Then change roles.

SELF-STUDY
AUDIO CD
CD-ROM

2 *Strategy plus* *Really?*

People say **Really?** to show
they are interested or surprised.

She's an artist.

Really?

In conversation . . .

Really is one of the top 50 words.

About you → **Pair work** Practice the conversations. Then ask the questions again.
Give your own answers.

1 A Where are you from?
B San Diego.
A San Diego? Really? I'm from Los Angeles.

2 A What's your name?
B Ryan.
A Really? My best friend's name is Ryan.

3 A Who's your best friend?
B Her name's Brittany.
A Really? What's she like?
B She's very nice.

3 *Listening and speaking* *Friends*

A Listen to these people talk about their friends. Write the missing words.

1 "Amy is a <u>friend</u> of mine
from the neighborhood.
She's about _____ years old.
Olivia is her _____.
She's the same age as my _____."

2 "Anton's a friend of mine.
He's my _____.
He's around my age.
He's a _____ guy – a fun guy."

3 "Gary is a friend from _____.
He's very _____.
His _____'s name is Gloria."

B *Pair work* Write the names of three people you know on a piece of paper.
Exchange lists. Ask questions about the people on your partner's list.

Chung Dae
Angela
Roberto

A Who's Chung Dae?
B He's my best friend.
A Really? Where's he from?

4 *Free talk* *Talk about your favorite people.*

See **Free talk 3** for more speaking practice.

Learning tip *Making diagrams*

Make diagrams with new vocabulary. An
example of a diagram is the family tree below.

1 Complete the family tree using the words in the box.

| grandmother | mother | sister | grandfather | brother | ✓father |

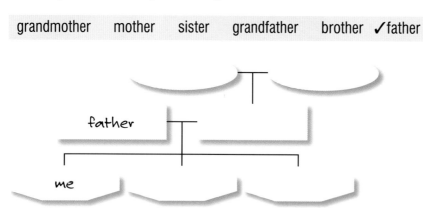

Mom or Mother?

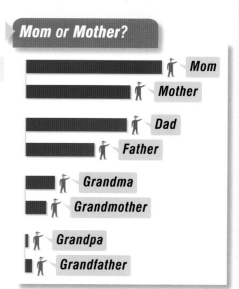

2 Now make your own family tree. Write notes about each person.

Her name's Hong.

She's seventy-two.

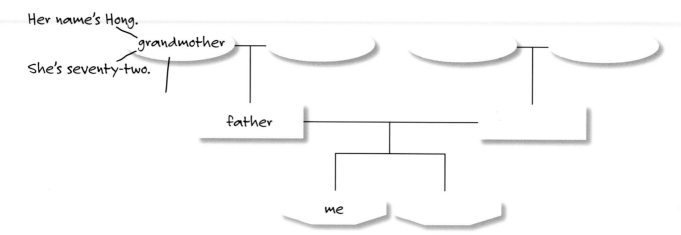

On your own

Make a photo album of your family and friends.
Write sentences about them in English.

This is my brother.
He's very smart.

1 Can you complete this conversation?

is
✓are
I'm
you're
he's
it's
we're
they're
his
her
my
your
our
their
this
these

Complete the conversation with the words in the box. Use capital letters when necessary. Then practice with a partner.

Angel Hi, Carla. How ___are___ you?

Carla _____ fine, thanks. Is _____ your car?

Angel No. _____ my brother's car. _____ on vacation.

Carla Cool. So, where is _____ brother?

Angel He and _____ wife are in Miami, with her parents. _____ family _____ from Miami, you know.

Carla Oh, right. So, are _____ children in Miami, too?

Angel No, _____ with my parents and me. _____ house is crazy. _____ all so busy with the kids.

Carla I bet _____ tired.

Angel Yeah, I really am. . . . Uh-oh, I'm late!

Carla OK. See you later.

Angel Wait! Where are my car keys? I mean, where are _____ brother's car keys?

Carla Are _____ his keys? Under the car? Here you go.

Angel Oh, thanks, Carla. You're wonderful!

2 Unscramble the questions.

Put the words in the correct order to make questions. Then ask and answer the questions with a partner.

1. full / teacher's / is / our / name / What ?
   <u>What is our teacher's full name ?</u>

2. phone / the / What's / number / school's ?
   _____

3. class / hard / our / English / Is ?
   _____

4. this class / students / Are / smart / the / in ?
   _____

5. today / not / Who's / in class ?
   _____

6. books / are / Where / your ?
   _____

3 How many words do you remember?

Complete the charts. Then make questions to ask and answer with a partner.

classroom items	locations in class	family and other people	words to describe people
clock	on the wall	neighbors	friendly

"Where's the clock?"   *"What's on the wall?"*

"What are your neighbors like?"   *"Are your neighbors friendly?"*

4 *Do you know these expressions?*

Complete the conversation with expressions from the box. Then practice with a partner.

Thank you.	Thanks anyway.	Nice to meet you.	✓Can I borrow your pen?	Really?
That's OK.	You're welcome.	Have a good day.	How do you spell *neighbor*?	

Anna Oh, no! Where's my pen? Excuse me. <u>Can I borrow your pen</u> ?

Michel Sure. Here you go.

Anna _____ .

Michel You're welcome.

Anna Hmm. _____ ?

Michel *Neighbor?* I'm sorry. I don't know.

Anna OK. _____ .

Michel Wait. Here's my dictionary.

Anna Oh, thanks.

Michel _____ .

Anna Oh. This is a French-English dictionary.

Michel Yes. I'm from France.

Anna France? _____ ? Uh-oh! My coffee! I'm sorry.

Michel _____ .

Anna By the way, I'm Anna.

Michel I'm Michel. _____ .

Anna Oh, no. I'm late for work. Sorry. Bye. _____ .

Michel Thanks. You too. Uh-oh. Where's my pen? And my dictionary?

5 *Who has the same answer?*

Class activity Complete the questions and write your answers.
Then ask your classmates the questions. Who has the same answer?

	Your answer	Classmates with the same answer
1. What's your mother<u>'s</u> first name?		
2. _____ old are your parents?		
3. _____ is your family from originally?		
4. _____ your best friend like?		
5. What's your best friend ____ name?		
6. _____ your favorite singer?		
7. _____ your favorite TV show?		

Self-check

How sure are you about these areas?
Circle the percentages.

grammar
20% 40% 60% 80% 100%

vocabulary
20% 40% 60% 80% 100%

conversation strategies
20% 40% 60% 80% 100%

· ·

Study plan

What do you want to review?
Circle the lessons.

grammar
1B 1C 2A 2B 2C 3A 3B 3C

vocabulary
1A 1C 2A 2B 2C 3A 3B 3C

conversation strategies
1D 2D 3D

Everyday life

In Unit 4, you learn how to . . .
- use simple present statements, *yes-no* questions, and short answers.
- talk about your daily and weekly routines.
- answer more than *yes* or *no* to be friendly.
- use *Well* to get time to think.

2

1

3

4

Before you begin . . .
Find these activities in the pictures. Which activities do you do every day? Check (✓) the boxes.

- ☐ do homework
- ☐ exercise
- ☐ work
- ☐ watch TV

In the morning

What's a typical morning like in your home?

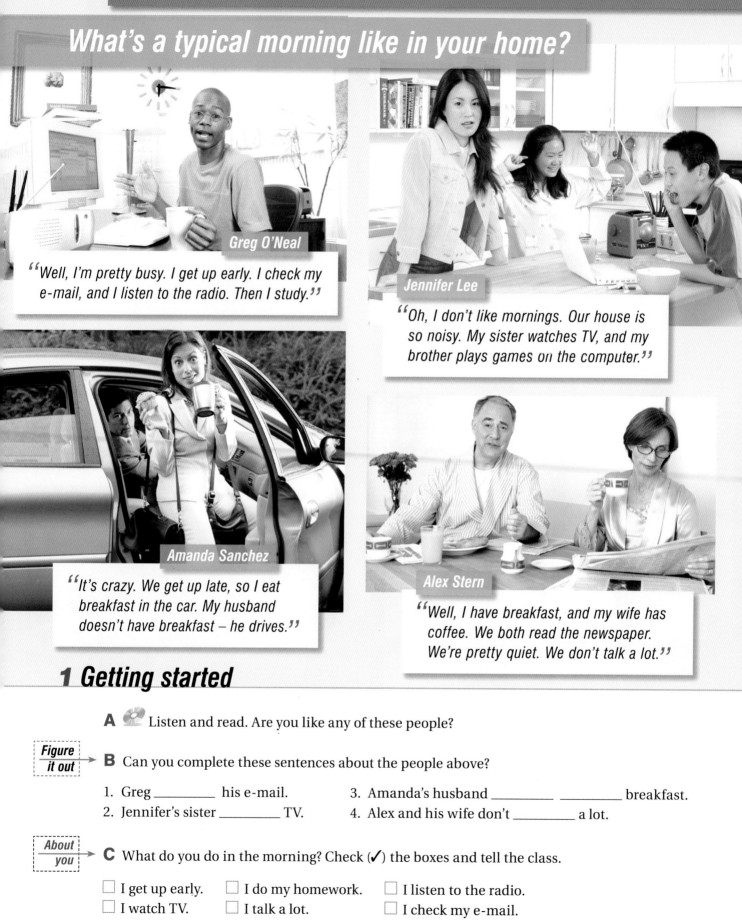

Greg O'Neal

"Well, I'm pretty busy. I get up early. I check my e-mail, and I listen to the radio. Then I study."

Jennifer Lee

"Oh, I don't like mornings. Our house is so noisy. My sister watches TV, and my brother plays games on the computer."

Amanda Sanchez

"It's crazy. We get up late, so I eat breakfast in the car. My husband doesn't have breakfast – he drives."

Alex Stern

"Well, I have breakfast, and my wife has coffee. We both read the newspaper. We're pretty quiet. We don't talk a lot."

1 Getting started

A 💿 Listen and read. Are you like any of these people?

Figure it out

B Can you complete these sentences about the people above?

1. Greg _____ his e-mail.
2. Jennifer's sister _____ TV.
3. Amanda's husband _____ _____ breakfast.
4. Alex and his wife don't _____ a lot.

About you

C What do you do in the morning? Check (✓) the boxes and tell the class.

- ☐ I get up early.
- ☐ I do my homework.
- ☐ I listen to the radio.
- ☐ I watch TV.
- ☐ I talk a lot.
- ☐ I check my e-mail.

2 Grammar *Simple present statements*

I	**eat**	breakfast.	I	**don't eat**	lunch.	**Verb endings: he, she, it**
You	**have**	coffee.	You	**don't have**	tea.	get → get**s**
We	**get up**	late.	We	**don't get up**	early.	watch → watch**es**
They	**read**	the paper.	They	**don't read**	books.	play → play**s**
He	**listens**	to the radio.	He	**doesn't listen**	to CDs.	study → stud**ies**
She	**watches**	TV.	She	**doesn't watch**	videos.	have → ha**s**

don't = do not   doesn't = does not   do → do**es**

A Complete these sentences.

1. I __don't like__ (not / like) mornings.
2. In my family, we _____ (have) breakfast together.
3. My mother _____ (not / watch) TV.
4. My father _____ (have) coffee.
5. My parents _____ (talk) a lot.
6. I _____ (not / read) the newspaper.
7. I _____ (check) my e-mail after breakfast.
8. My best friend _____ (not / get up) early in the morning.

In conversation . . .

Don't and **doesn't** are more common than **do not** and **does not**.

don't

do not

doesn't

does not

About you → **B** Now write four sentences about your mornings. Compare with a partner.

I don't eat breakfast.

➡ A **I don't eat breakfast. How about you?**
B **I have breakfast every morning.**

3 Speaking naturally *-s endings of verbs*

/s/ like**s**   /z/ listen**s**   /ɪz/ relax**es**

A  Listen and repeat the words above. Notice the verb endings.

B Listen to the questions. Do the verbs end in /s/, /z/, or /ɪz/?

In your group . . .	/s/	/z/	/ɪz/
1. Who uses an alarm clock?	☐	☐	✔
2. Who gets up late?	☐	☐	☐
3. Who exercises in the morning?	☐	☐	☐
4. Who sings in the shower?	☐	☐	☐
5. Who eats a big breakfast?	☐	☐	☐
6. Who drives to class?	☐	☐	☐

About you → **C Group work** Take turns asking and answering the questions.

"Who uses an alarm clock?"   *"I use an alarm clock."*

1 Building vocabulary

A Listen and say the expressions. Then check (✓) the things you do every week. Can you add more activities?

- [ ] take a class
- [ ] play sports
- [ ] clean the house
- [ ] go shopping
- [ ] do the laundry
- [ ] make phone calls

Word sort

B For each day of the week, think of one thing you usually do. Then tell the class.

Sunday	Monday	Tuesday	Wednesday	Thursday	Friday	Saturday
	play soccer					

"On Mondays, I play soccer."

2 Building language

A Look at the questionnaire. Can you complete the last two questions? Then listen and check (✓) the answers that are true for you.

Do you have a weekly routine?	Yes, I do.	No, I don't.
1. Do you play sports every week?	☐	☐
2. Do you take any lessons or classes?	☐	☐
3. Do you check your e-mail on the weekends?	☐	☐
4. Do you make a lot of phone calls on Saturdays?	☐	☐
5. _____ you _____ shopping on Sundays?	☐	☐
6. _____ you _____ the laundry every week?	☐	☐

Figure it out

About you

B Pair work Ask and answer all the questions. Can you give more information?

"Do you play sports every week?"   *"Yes, I do. I play tennis on Saturdays."*

3 Grammar Yes-No questions and short answers

Do you **go** to a class in the evening? Yes, I **do**. / No, I **don't**.	**Do** you and your friends **play** sports after class? Yes, we **do**. / No, we **don't**.
Does your mother **work** on the weekends? Yes, she **does**. / No, she **doesn't**.	**Do** your friends **make** phone calls at night? Yes, they **do**. / No, they **don't**.

A Complete the questions. Compare with a partner.

1. _Do_ you eat a lot of snacks every day?
2. ＿＿ you make a lot of phone calls before breakfast?
3. ＿＿ you clean the house on the weekends?
4. ＿＿ you ＿＿＿＿ your homework late at night?
5. ＿＿ you ＿＿＿＿ TV after dinner?
6. ＿＿ your friends ＿＿＿＿ their e-mail every day?
7. ＿＿ your teacher work in the evening?
8. ＿＿ your best friend ＿＿＿＿ a class on Saturdays?

> **Time expressions**
>
> **on** Monday(s)
> **on** (the) weekends
> **on** the weekend
> **in** the morning(s)
> **in** the afternoon(s)
> **in** the evening(s)
> **at** night
> **before** breakfast
> **after** class
> **every** day

About you → **B** *Pair work* Ask and answer the questions.
How many of your answers are the same?

A Do you eat a lot of snacks every day?
B Yes, I do. I eat two or three snacks in the afternoon.

4 Survey

A Find people who do these things. Write their names in the chart.

Who has a busy week?

Find someone who . . .	Name
belongs to a club.	＿＿＿＿＿
plays on a team.	＿＿＿＿＿
works on the weekends.	＿＿＿＿＿
has breakfast in the car.	＿＿＿＿＿
studies English before breakfast.	＿＿＿＿＿
gets up early on Sundays.	＿＿＿＿＿

"Do you belong to a club?"   *"Yes, I do. I belong to a chess club."*

B Tell the class something interesting about a classmate.

"Anton belongs to a chess club."

5 Vocabulary notebook *Verbs, verbs, verbs*

See page 42 for a new way to log and learn vocabulary.

1 Conversation strategy *Saying more than yes or no*

A Can you answer this question with more than *yes* or *no*?

A *Do you live around here?*

B *Yes, _____ .* **or** *No, _____ .*

Now listen. What do you find out about Ray?

Tina Hi. I see you here all the time. Do you come here every day?

Ray No. . . . Well, I have breakfast here before class.

Tina Oh, are you a student?

Ray Yes. I'm a law student.

Tina Really? I'm in the business school.

Ray Oh. So do you live around here?

Tina Well, I live about 20 miles away, in Laguna Beach.

Ray So, are you from California?

Tina Well, I'm from Chicago originally, but my family lives here now.

Notice how Ray answers Tina's questions. He says more than *yes* or *no*. He wants to be friendly. Find examples in the conversation.

"Oh, are you a student?"

"Yes. I'm a law student."

About you

B Match the questions and answers. Then ask and answer the questions with a partner. Give your own answers, saying more than *yes* or *no*.

1. Do you live around here? _f_
2. Are you from here originally? _____
3. Do you have a part-time job? _____
4. Do you like sports? _____
5. Do you have brothers and sisters? _____
6. Do your parents live around here? _____

a. Yeah. Well, I play on a softball team.
b. Yeah, I work at a restaurant on the weekends.
c. No, they live in a small town near the ocean.
d. No, I'm from Rio originally.
e. No, I'm an only child.
f. No, I live near the beach.

SELF-STUDY
AUDIO CD
CD-ROM

2 *Strategy plus* *Well*

Start your answer with *Well* if you need time to think, or if your answer is not a simple *yes* or *no*.

Are you from California?

Well, I'm from Chicago originally, . . .

In conversation . . .

Well is one of the top 50 words.

About you → *Pair work* Practice the conversations. Then ask the questions again. Give your own answers.

1 *A* What are your neighbors like?
   B Well, they're very noisy. They like loud music.

2 *A* Do you see your family a lot?
   B Well, not really. They don't live around here.

3 *A* Do you study every day?
   B Well, not every day. I go out with friends on the weekends.

3 *Listening and speaking* *What's the question?*

A Listen to people answer the questions below. Which question is each person answering? Number the questions.

☐ *"Do you go out on the weekends?"*	
[1] *"Do you read a lot?"*	☐ *"Do you know people from other countries?"*

☐ *"Do you live with your parents?"* ☐ *"Do you exercise every day?"*

About you → **B** *Pair work* Ask and answer the questions above. Be sure to say more than *yes* or *no* in your answers. Use *Well* if you need to.

C Look at each question again. Change the verb. How many new questions can you make? Then ask a partner your questions.

Do you ~~go out~~ on the weekends?
   read
   watch TV

On average . . .

1 Reading

A How much time do you spend on these activities every day? Tell the class.

- on the phone _two hours_
- in bed _____
- at work or at school _____
- in the car, or on the bus or train _____

"I spend two hours a day on the phone."

B Read the article. How many daily activities does it talk about?

IN THE LIFETIME
OF AN AVERAGE AMERICAN...

SLEEP-O-METER
1 9 8 6 4 5 HOURS
3 6 MINUTES

How many hours do you spend in bed? Six or seven hours a night maybe? And how many hours do you spend in front of the TV every week? Nine or ten? That's not a lot, is it? Well, think again. Add together all the hours you spend on these activities in a lifetime, and the total numbers are surprising.

In an average lifetime, an American works over 90,000 hours, walks an amazing 22,000 kilometers (14,000 miles), and spends three and a half years eating.

Do you call your friends a lot? An average American talks on the telephone for two and a half years. On average, Americans sleep for 24 years and watch TV for 12 years. That's 36 years – about half a lifetime – in bed or on the couch!

C Read the article again, and complete the sentences. Compare answers with a partner. Are any facts surprising?

In a lifetime, an average American spends . . .
1. _____ hours at work.
2. _____ years on the telephone.
3. _____ years in bed.
4. _____ years in front of the TV.

2 *Listening* Teen habits

A Read about the habits of an average American teenager. Then listen to Christine talk about her habits. Complete the chart about Christine.

An average teenager . . .

drinks 16 cans of soda a week.
eats dinner at home 3 times a week.
spends about 5 hours a month online.
watches TV 20 hours a week.

Christine . . .

drinks about _____ cans of soda a week.
eats dinner at home _____ times a week.
spends about _____ hours a month online.
watches TV _____ hours a week.

About you → **B** *Pair work* Do you have the same habits as an average American teenager? Ask and answer questions.

"Do you drink 16 cans of soda a week?"   *"No, I drink about 2 cans a week. What about you?"*

3 *Writing and speaking* An average week

A Complete the sentences. Then tell a partner. Take notes on your partner's activities.

My week: On average . . .	**My partner's week: On average . . .**
I study / work _____ hour(s) a week.	_____
I exercise _____ hour(s) a week.	_____
I use a cell phone _____ time(s) a day.	_____
I go out with my friends _____ night(s) a week.	_____
I spend _____ hour(s) with my family on weekends.	_____

B Write an e-mail message to a friend about your partner. Use your notes to help you.

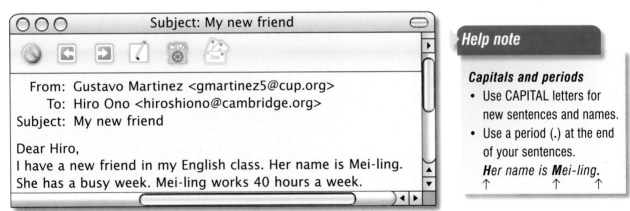

Subject: My new friend

From: Gustavo Martinez <gmartinez5@cup.org>
To: Hiro Ono <hiroshiono@cambridge.org>
Subject: My new friend

Dear Hiro,
I have a new friend in my English class. Her name is Mei-ling.
She has a busy week. Mei-ling works 40 hours a week.

Help note

Capitals and periods
• Use CAPITAL letters for new sentences and names.
• Use a period (.) at the end of your sentences.
Her name is Mei-ling.
↑   ↑   ↑

C *Group work* Take turns reading your messages aloud. Who has a different or surprising routine?

4 *Free talk* Interesting facts

See *Free talk 4* for more speaking practice.

Learning tip *Drawing pictures*

Draw and label simple pictures in your notebook. The pictures below show different verbs.

1 Label the pictures. Use a verb to describe each activity.

  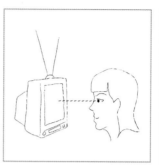 

read the newspaper

2 Draw and label your own pictures of activities.

3 Complete the chart with your everyday activities.

Write two things you do . . .		
every day	I read the newspaper.	
in the afternoon		
on Sundays		
after breakfast		
before bed		

On your own

Write labels for the things you do every day.
Put your labels around the house.

Free time

In Unit 5, you learn how to . . .

- ask simple present information questions.
- say how often you do things.
- talk about free-time activities and TV shows.
- ask questions in two ways.
- use *I mean*.

Before you begin . . .

Do you do these things on the weekends?
Check (✓) the boxes.

- ☐ go on the Internet
- ☐ go to a club
- ☐ eat out
- ☐ go to the movies

FREE-TIME SURVEY

Name: Robert Acosta **Occupation:** Student

Please answer the questions.
You can check (✓) more than one answer.

1. **How often do you go out?**
   - ☐ every night
   - ✓ once or twice a week
   - ☐ other _____

2. **When do you usually go out?**
   - ☐ on weeknights
   - ☐ on weekends

3. **Where do you go?**
   - ☐ to the movies
   - ☐ to clubs
   - ☐ to restaurants
   - ☐ to the gym
   - ☐ other _____

4. **Who do you go out with?**
   - ☐ my family
   - ☐ my friends
   - ☐ my girlfriend / boyfriend
   - ☐ other _____

5. **How often do you eat out?**
   - ☐ every night
   - ☐ once or twice a week
   - ☐ two or three times a month
   - ☐ other _____

6. **What do you do in your free time at home?**
   - ☐ go on the Internet
   - ☐ relax in front of the TV
   - ☐ rent movies
   - ☐ other _____

Thank you for your help with our survey!

1 Getting started

A 💿 Listen and read as Robert completes the survey with a friend. Check (✓) his answers.

Figure it out

B Can you complete these questions? Use the survey to help you. Compare with a partner.

❶ *A* _____ do you go to the movies?
B Once or twice a month.

❷ *A* _____ do you go to the movies with?
B My best friend.

❸ *A* _____ do you go with your friends?
B To restaurants and clubs.

About you

C *Pair work* Complete the survey for your partner. Ask and answer the questions.

44

2 *Grammar* *Simple present: Information questions*

What	**do** you	**do** in your free time?	Meet my friends.
Who	**do** you	**go out** with?	A friend.
Where	**does** she	**go**?	To the movies.
How often	**does** he	**eat out**?	Twice a month.
When	**do** they	**go out**?	On the weekends.

▶**Time expressions**

How often?
every night
on Friday nights
once a week
three times a week
twice a month

A Unscramble the words to make questions. Compare with a partner.

1. do / you / do / what / on Friday nights ?

   <u>What do you do on Friday nights?</u>

2. after class / where / go / your friends / do ?

   _____

3. you / who / do / on the weekends / go out with ?

   _____

4. do / your parents / how often / go on the Internet ?

   _____

5. your family / does / have dinner together / when ?

   _____

6. on weeknights / go out / do / you / how often ?

   _____

About you ➡ **B** *Pair work* Ask and answer the questions with a partner.

A **What do you do on Friday nights?**
B **I go to a club.**

3 *Speaking naturally* *Do you . . . ?*

Do you go out a lot?   Where **do you** go?   What **do you** do?

A Listen and repeat the questions above. Notice the pronunciation of *do you*.

B Listen to the conversations. Write the questions you hear.

① *A* <u>Do you relax in your free time?</u>
   B Well, yes, on the weekends.
   A _____
   B I sleep late, read, watch TV . . .

② *A* _____
   B Yes, I do. I like movies a lot.
   A _____
   B Two or three times a week.

About you ➡ **C** *Pair work* Practice the conversations. Then ask and answer the questions.
Give your own answers.

TV shows

1 Building language

A Listen. When does Marisa watch TV? Practice the conversation.

Steve This soup is delicious. What's in it? . . . Marisa? Marisa!

Marisa I'm sorry, what?

Steve You know, sometimes I think you watch too much TV.

Marisa Oh, I hardly ever watch TV.

Steve Are you serious?

Marisa Well, sometimes I watch the morning shows.
And I usually watch the late movie.

Steve And you always have dinner in front of the TV!
I mean, you never talk to me.

Marisa Yes, I do! I talk to you during the commercials.

Figure it out

B Can you find words to complete the sentences?
Use the conversation to help you.

1. _____ Marisa watches the morning shows.
2. Marisa _____ watches the late movie.
3. Marisa _____ has dinner in front of the TV.

2 Grammar *Frequency adverbs*

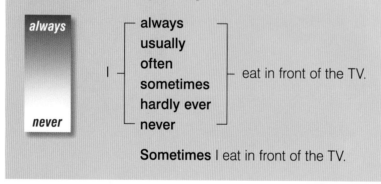

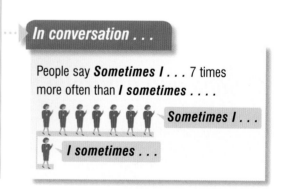

In conversation . . .

People say **Sometimes I** . . . 7 times more often than **I sometimes** . . . .

Sometimes I . . .

I sometimes . . .

About you

Add frequency adverbs to make true sentences.
Then compare with a partner.

1. I *never* watch TV in the morning.
2. I watch TV shows in English.
3. My family has dinner in front of the TV.
4. I rent movies on the weekends.
5. My family watches TV late at night.
6. We watch videos in our English class.

A I never watch TV in the morning.
B Really? I always watch TV in the morning.

3 *Building vocabulary*

A 🔊 Listen. What kinds of TV shows do you hear? Write the number next to the type of show.

☐ cartoon	☐ soap opera	☐ talk show	☐ game show
☐ documentary	☐ reality show	1 sitcom	☐ the news

Word sort ➤ **B** What kinds of shows do you like and dislike? Complete the chart. Add other kinds of shows you know.

Likes

☺☺☺ = I love cartoons _____
☺☺ = I really like _____
☺ = I like _____

Dislikes

☹☹☹ = I hate _____
☹☹ = I can't stand _____
☹ = I don't like _____

About you ➤ **C** *Pair work* Find out what kinds of TV shows your partner likes.

"Do you like cartoons?"   *"Yes, I do. I love cartoons. My favorite is . . ."*

4 *Talk about it* Do you watch too much TV?

Group work Discuss the questions. Do you have the same TV-watching habits?

▸ How many TVs do you have at home?
▸ How often do you watch TV?
▸ Do you have breakfast in front of the TV?

▸ Do you ever watch TV in bed? in restaurants?
▸ Do you watch the commercials on TV?
▸ Do you think you watch too much TV?

5 *Vocabulary notebook* Do what? Go where?

See page 52 for a new way to log and learn vocabulary.

Do you go straight home?

1 Conversation strategy *Asking questions in two ways*

A Can you complete the second question?

    A **What do you do after work? Do you _____ ?**
    B **Well, I usually go shopping and then go home.**

Now listen. What does Lori do after class?

Adam **So, what do you do after class? Do you go straight home?**

Lori **Well, usually. Sometimes I meet a friend for dinner.**

Adam **Oh, where do you go? I mean, do you go somewhere nice?**

Lori **Do you know Fabio's? It's OK. I mean, the food's good, and it's cheap, but the service is terrible. Do you know it?**

Adam **Well, actually, I work there. I'm a server.**

Notice how Adam asks questions in two ways. His questions are clear and not too direct. Find examples in the conversation.

"So, what do you do after class? Do you go straight home?"

B Match the first question to a good second question.

1. What do you do after class? _c_
2. How do you get home? ____
3. Do you ever feel tired after class? ____
4. Do you work in the evening? ____
5. How often do you go shopping? ____
6. What do you do for lunch? ____

a. I mean, do you eat out?
b. Do you go shopping a lot?
c. Do you go out for coffee?
d. I mean, do you usually need a break?
e. Do you take the subway or the bus?
f. I mean, do you have a part-time job?

About you → **C Pair work** Ask and answer the pairs of questions. Give your own answers.

"What do you do after class? Do you go out for coffee?"   *"Well, I usually . . ."*

SELF-STUDY
AUDIO CD
CD-ROM

2 Strategy plus *I mean*

You can use *I mean*
to repeat your ideas or to
say more about something.

Where do you go?
I mean, do you go
somewhere nice?

Do you know Fabio's?
It's OK. I mean, the
food's good, . . .

In conversation . . .

I mean is one of the top 15 expressions.

A Complete the questions or answers with your own ideas.
Compare with a partner. Do you have any of the same ideas?

1. *A* Do you ever go out after class?
   B Well, not very often. I mean, I usually go _straight home_ .

2. *A* How do you like the restaurants in your neighborhood?
   B They're not bad. I mean, they're _____ .

3. *A* Are you busy in the evening? I mean, do you _____ ?
   B Well, I take a lot of classes.

4. *A* What do you do in your free time?
   B Well, I don't have a lot of free time. I mean, _____ .

> About
> you

B *Pair work* Ask and answer the questions. Give your own answers.

3 Listening and speaking *What do they say next?*

A Listen to the beginning of three conversations. How do you think
each conversation continues? Circle *a* or *b*.

Conversation 1
a . . . what are your hobbies?
b . . . where do you work?

Conversation 2
a . . . I take French, too.
b . . . the food is good.

Conversation 3
a . . . do you watch TV?
b . . . do you live around here?

B Now listen to the complete conversations. Check your answers.

> About
> you

C Add a second question to each question below. Then choose one and start
a conversation with a partner.

1. How often do you play sports? I mean, do you play _____ ?
2. Where do you usually have dinner? I mean, do you eat _____ ?
3. What do you do on the weekends? I mean, do you _____ ?

4 Free talk *Play a board game.*

See *Free talk 5* for more speaking practice.

Internet addicts

1 Reading

A Check (✓) the statements you agree with. Compare with a partner. Can you add more ideas?

> **The Internet is a great place to . . .**
> ☐ make new friends and "chat." ☐ find information.
> ☐ spend your free time. ☐ practice your English.
> ☐ listen to music.

B Read the article. Do you know any Internet "addicts"?

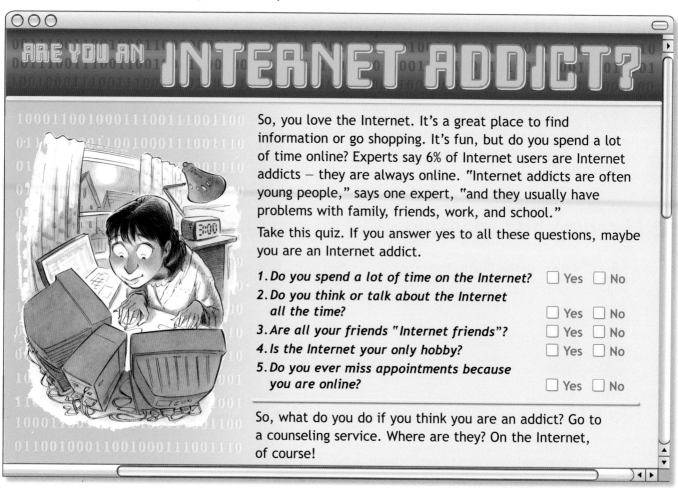

ARE YOU AN INTERNET ADDICT?

So, you love the Internet. It's a great place to find information or go shopping. It's fun, but do you spend a lot of time online? Experts say 6% of Internet users are Internet addicts — they are always online. "Internet addicts are often young people," says one expert, "and they usually have problems with family, friends, work, and school."

Take this quiz. If you answer yes to all these questions, maybe you are an Internet addict.

1. Do you spend a lot of time on the Internet? ☐ Yes ☐ No
2. Do you think or talk about the Internet all the time? ☐ Yes ☐ No
3. Are all your friends "Internet friends"? ☐ Yes ☐ No
4. Is the Internet your only hobby? ☐ Yes ☐ No
5. Do you ever miss appointments because you are online? ☐ Yes ☐ No

So, what do you do if you think you are an addict? Go to a counseling service. Where are they? On the Internet, of course!

C Answer the questions about the article. Compare your answers with a partner.

1. How many Internet users are "addicts"?
2. What problems do Internet addicts have?
3. Where do Internet addicts go for help?
4. What are some things Internet addicts do?

About you → **D** *Pair work* Take the quiz in the article. Ask and answer the questions. Is your partner an Internet addict? Are you?

2 *Listening and speaking* *Using computers*

A Why do people use computers? How many different ideas can you think of?

"They watch DVDs. They . . ."

B  Listen. What do Andrea and Yoshi use their computers for?
Check (✓) the boxes.

Andrea

☐ *She watches DVDs.*
☐ *She plays CDs.*
☐ *She checks her e-mail.*
☐ *She has a Web site.*

Yoshi

☐ *He practices English.*
☐ *He looks at digital photos.*
☐ *He buys books online.*
☐ *He pays bills online.*

About you

C *Group work* Discuss the questions. Do you use computers for the same things?

■ How often do you use a computer?
What do you use it for?

■ Do you go on the Internet?
What do you do online?

■ Do you have e-mail?
How often do you send e-mail?

■ Do you ever shop online?
What do you buy?

3 *Writing* *A message to a Web site*

A Write a message to the Web site about yourself. Complete the sentences.

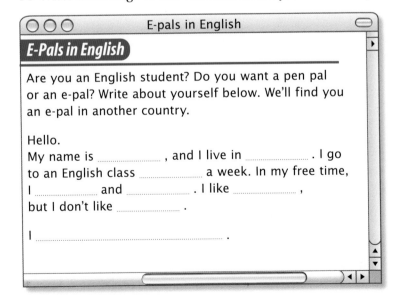

◯◯◯ E-pals in English

E-Pals in English

Are you an English student? Do you want a pen pal
or an e-pal? Write about yourself below. We'll find you
an e-pal in another country.

Hello.
My name is .... , and I live in .... I go
to an English class .... a week. In my free time,
I .... and .... I like .... ,
but I don't like ....

I ....

Help note

Linking ideas with and and but

*My name is Sombat, **and** I live in Bangkok.*
*I take English **and** Chinese.*
*I love movies, **but** I don't like cartoons.*

B *Class activity* Read your classmates' messages. Choose an e-pal and tell the class
about him or her.

Learning tip *Verbs + . . .*

Write down verbs and the words you can use *after* them.

play	music
	sports
	soccer

1 Which words and expressions in the box go with the verbs below? Complete the chart.

breakfast	the laundry	homework		lessons	✓ music	snacks
a class	dinner	computer games		meals	on a team	soccer

play	music	**eat**		**take**		**do**	

2 Now think of words and expressions that go with these verbs.

go	to a class	**watch**	documentaries	**read**	
	out				

On your own

Make a vocabulary "flip pad." On each page, write a verb with words you can use after it. Look through it when you have time.

Neighborhoods

In Unit 6, you learn how to . . .

- use *There's* and *There are*.
- use *some*, *no*, *a lot of*, and *a couple of*.
- talk about your neighborhood and local events.
- ask for and tell the time.
- use *Me too* or *Me neither* to show you're like someone.
- use *Right* and *I know* to agree.

The Mall of America, Bloomington, Minnesota

Yankee Stadium, New York City

Stanley Park, Vancouver

The J. Paul Getty Museum, Los Angeles

Before you begin . . .

Do you have any places like these in your city or town?

How often do you go to them?

THE DAILY HERALD

How do you like your neighborhood?

People talk about the popular neighborhood called Parkview.

" Well, Parkview is convenient. There's a big supermarket and some nice stores, but there's no mall. We need a mall! **"**

– Janet Carson, 47, medical researcher

" Um, it's nice. There are two nice outdoor cafés and a couple of movie theaters. There's a new swimming pool in the park – we have a beautiful little park. Yeah, it's good. **"**

– Rick Martinez, 33, stockbroker

" Parkview is boring! There's no place to go. I mean, there's no mall, no fast-food places – just a lot of expensive restaurants. Oh, and a small park. **"**

– Megan Novak, 15, high school student

1 Getting started

A Listen and read. Which people like Parkview? Why?

Figure it out → **B** What's in your neighborhood? Circle the words to make true sentences. Compare with a partner.

1. There's **a / no** mall.
2. There are **no / some / a lot of** cheap restaurants.
3. There are **no / a couple of / some** movie theaters.

2 Grammar *There's and There are; quantifiers*

There's **a** park.	**Adjectives before nouns**
There's **no** mall.	There's a **small** park.
There are **a lot of** restaurant**s**.	There's a **beautiful** pool.
There are **some** outdoor café**s**.	There's a **new** restaurant.
There are **a couple of** movie theater**s**.	There are some **expensive** stores.
There are **no** club**s**.	

There's = There is

▶ **In conversation . . .**

People often say **There's** before plural nouns, but it is not correct to write this.

A These sentences about the neighborhood on page 54 aren't accurate. Can you correct them? Then compare with a partner.

1. There are a couple of big supermarkets.   <u>There's a big supermarket.</u>

2. There are no swimming pools.   _____

3. There's a big park.   _____

4. There's one movie theater.   _____

5. There's an expensive restaurant.   _____

6. There are a lot of fast-food places.   _____

7. There are a lot of apartment buildings.   _____

8. There's a stadium.   _____

About you → **B** *Pair work* Student A: Say what's in your neighborhood.
Student B: Ask for more information. Then change roles.

A **There's a big stadium in my neighborhood.**

B **What sports do they play there?** **or** **How often do you go there?**

3 Speaking naturally *Word stress*

● ●	● ● ●	● ● ●
*mo*vie	*sta*dium	a*part*ment

A Listen and repeat the words above. Notice the word stress.

B Listen. Write the words in the correct column.

✓ movie	museum
✓ stadium	neighborhood
✓ apartment	music
expensive	beautiful
noisy	boring
theater	convenient

❶ ● ●	❷ ● ● ●	❸ ● ● ●
movie	stadium	apartment

C *Group work* Talk about a perfect neighborhood. What's there? What's not there? Agree on a list of places. Then tell the class.

"In a perfect neighborhood, there's a beautiful park."   *"And there are some cafés. . . ."*

1 Building vocabulary

A Listen and say the times. What time is it now?

It's eleven (**o'clock**).

It's two-oh-five.
It's five **after** two.

It's four-fifteen.
It's **a quarter after** four.

It's ten-thirty.

It's six-forty-five.
It's **a quarter to** seven.

It's eight-fifty.
It's ten **to** nine.

It's twelve a.m.
It's **midnight**.

It's twelve p.m.
It's **noon**.

Notice . . .

a.m. = **before** 12 noon
p.m. = **after** 12 noon

B *Pair work* Take turns asking and telling the time.

In conversation . . .

People say (hour)-*fifteen* more
than *a quarter after* (hour).

two-fifteen

a quarter after two

"*What time is it?*"   "*It's two-fifteen.*" **or** "*It's a quarter after two.*"

2 Listening *What's on this weekend?*

How often do you go to events like these? Tell the class. Then listen to
the radio show, and complete the chart.

Event	Where is it?	What time does it start?
1. concert		
2. art exhibit		
3. soccer match		
4. play		

3 Building language

A Listen. What time is the concert? Practice the conversation.

Kyle Hey, there's a free jazz concert tomorrow night.

Erin Oh, that sounds like fun. Where?

Kyle At Grant Park.

Erin What time does it start?

Kyle Um, it starts at . . . 7:00.

Erin OK, well, let's go. Let's meet at the park at a quarter to seven.

Kyle But they don't usually have a lot of seats, so . . .

Erin Oh, well, in that case, let's get there early – say, around 6:30.

Figure it out

B Complete these suggestions with verbs. Have a conversation with a partner.

1. Let's _____ to a movie tomorrow.
2. Let's _____ coffee together after class.

4 Grammar *Telling time; suggestions with Let's*

What time is it?	**It's** 6:30.	**Suggestions**
What time does the concert start?	It starts **at** 9 o'clock.	**Let's go.**
What time do supermarkets close?	**(At) about** 10:00 p.m.	**Let's meet** at 6:45.
What time do you go out at night?	Usually **around** 8:00 or 8:30.	**Let's get** there early.

A Write questions with *What time.* Then ask three classmates your questions.

> **In conversation . . .**
>
> You can ask people you don't know
> ***Excuse me, do you have the time?***

Find out what time . . .

1. they get home on Saturday nights.   <u>*What time do you get home on Saturday nights?*</u>
2. they leave work or school in the afternoon. _____
3. their local supermarket opens and closes. _____
4. their favorite TV show starts. _____
5. the last train leaves their local station. _____
6. buses start running in the mornings. _____

About you

B *Pair work* Talk about three events this week. Make plans to go to an event together. Use the conversation in 3 above to help you.

5 Vocabulary notebook *A time and a place . . .*

See page 62 for a new way to log and learn vocabulary.

It's a great place to live.

1 Conversation strategy *Me too* and *Me neither*

A Can you match each statement with the correct response?

1. I love our neighborhood. _____
2. I don't like the new movie theater. _____

a. Me neither.
b. Me too.

Now listen. What do you find out about this neighborhood?

Ben	I just love this neighborhood.
Jessica	Me too. I bet it's a great place to live.
Ben	Yeah. It has some great restaurants.
Jessica	Right. But they're expensive.
Ben	Yeah, I know. There are a lot of rich people around here.
Jessica	Well, I'm not rich!
Ben	No, me neither.
Jessica	By the way, are you hungry? I'm starving.
Ben	Me too. But let's eat somewhere else. It's kind of expensive around here.

Notice how Ben and Jessica say *Me too* and *Me neither* to show they have something in common. Find the examples in the conversation.

"I just love this neighborhood."
"Me too."

About you → **B** Make true sentences about your neighborhood. Circle an expression or add your own.

1. I live in **an exciting** / **a boring** / __a great__ neighborhood.
2. I like the **stores** / **houses** / _____ in my neighborhood.
3. I don't like the **restaurants** / **buildings** / _____ there.
4. I go to a lot of **movies** / **concerts** / _____ in my neighborhood.
5. I don't **go shopping** / **eat out** / _____ there.

C *Group work* Read your sentences aloud. Who has something in common with you? Find someone who answers *Me too* or *Me neither.*

"I live in an exciting neighborhood."   "Me too." **or** "Oh, really? I live in a boring neighborhood."

SELF-STUDY
AUDIO CD
CD-ROM

2 Strategy plus *Right and I know*

Say **Right** and **I know** to show you agree
with someone, or that you are listening.

In conversation . . .

Right is one of the top 50 words,
and **know** is one of the top 20.

It has some great restaurants.

Right. But they're expensive.

Yeah, I know.

About you → Complete the statements with your own ideas. Then practice with a partner.
Respond with *Right* and *I know.*

1. A Every neighborhood needs a ___stadium___ .
   B _____ .

2. A The best neighborhood in town is _____ .
   B _____ .

3. A There are no good _____ around here.
   B _____ .

4. A I don't like the _____ .
   B _____ .

5. A A lot of rich people live in _____ .
   B _____ .

3 Listening *City living*

A  Listen to Sam talk about his neighborhood. What topics is he talking
about? Circle *a* or *b*.

❶
  a *the restaurants*
  b *the people*

❷
  a *his neighbors*
  b *places to shop*

❸
  a *concerts*
  b *sporting events*

B Now listen to three things Sam says. Decide if you are like Sam or different
from Sam. Complete a response in the chart.

I'm like Sam.	I'm different from Sam.
❶ *Me too. My neighborhood _____ .*	*Really? I live _____ .*
❷ *I know. I like _____ .*	*Yeah? I like _____ .*
❸ *Me neither. I don't like _____ .*	*Really? I like _____ .*

4 Free talk *Find the differences.*

See **Free talk 6** for more speaking practice.

1 Reading

A What classified ads do you find in a local newspaper? Look at these headings. Add your ideas below.

Classifieds

CLASSES	ITEMS FOR SALE	HELP WANTED	LOCAL EVENTS
piano lessons	cars	baby-sitters	concerts

B Read these classified ads. Choose one of the headings above for each one.

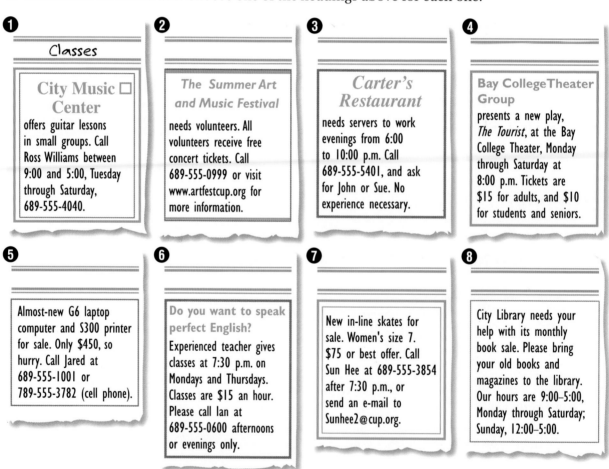

❶

Classes

City Music □ Center
offers guitar lessons in small groups. Call Ross Williams between 9:00 and 5:00, Tuesday through Saturday, 689-555-4040.

❷

The Summer Art and Music Festival
needs volunteers. All volunteers receive free concert tickets. Call 689-555-0999 or visit www.artfestcup.org for more information.

❸

Carter's Restaurant
needs servers to work evenings from 6:00 to 10:00 p.m. Call 689-555-5401, and ask for John or Sue. No experience necessary.

❹

Bay College Theater Group
presents a new play, *The Tourist*, at the Bay College Theater, Monday through Saturday at 8:00 p.m. Tickets are $15 for adults, and $10 for students and seniors.

❺

Almost-new G6 laptop computer and $300 printer for sale. Only $450, so hurry. Call Jared at 689-555-1001 or 789-555-3782 (cell phone).

❻

Do you want to speak perfect English?
Experienced teacher gives classes at 7:30 p.m. on Mondays and Thursdays. Classes are $15 an hour. Please call Ian at 689-555-0600 afternoons or evenings only.

❼

New in-line skates for sale. Women's size 7. $75 or best offer. Call Sun Hee at 689-555-3854 after 7:30 p.m., or send an e-mail to Sunhee2@cup.org.

❽

City Library needs your help with its monthly book sale. Please bring your old books and magazines to the library. Our hours are 9:00–5:00, Monday through Saturday; Sunday, 12:00–5:00.

C Find and circle the following information in the ads. Then compare your answers with a partner.

- the telephone number for the guitar lessons
- the cost of the English classes
- three items for sale
- the time the play starts
- three words you want to learn
- an ad that sounds interesting

2 *Talk about it* *Too much advertising?*

Group work How often do you see ads like these? Do you think there's too much advertising around you? Discuss the questions.

▶ Do you ever read **pop-up ads** on the Internet?

▶ Do you get a lot of **"spam"** – unwanted e-mails?

▶ What do the **billboards** in your city advertise?

▶ What do you think of the **commercials** on TV?

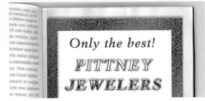

▶ Do you think there's too much advertising in **newspapers** and **magazines**?

▶ Do you ever see people in funny **costumes**? What do they advertise?

3 *Writing and speaking* *Bulletin boards*

A Do you ever read notices or ads on a bulletin board? Which ones do you read?

B Write an ad for a bulletin board. Use one of the ideas below.

Help note

Prepositions

Store hours are **from** 6:00 **to** 10:00.
Call **between** 9:00 **and** 5:00.
The store is open Monday **through** Saturday.
The play is **at** 8 p.m. **at** the library.
Call Jim **at** 555-7777, or **on** his cell phone.
Call us **for** more information.

C **Group work** Take turns reading your ads aloud. Ask questions to find out more information.

A **Do you need yoga classes?** . . .
B **What time do your yoga classes finish?** . . .
C **Are your classes fun?** . . .

1 When do you usually do these things each day? Write the times.

6:30 a.m.	get up			finish work / school
	eat breakfast			get home
	go to work / school			have dinner
	eat lunch			go to bed

2 Complete the daily planner. What do you (and your family) usually do at different times?

6:00–9:00 a.m.

9:00 a.m.–12:00 p.m.

12:00–2:00 p.m.

2:00–5:00 p.m.

5:00–8:00 p.m.

8:00 p.m. –12:00 a.m.

On your own

Draw a clock face. Where are you at each time of the day? Write notes.

1 That's not quite right.

Which of these sentences are true for you? Check (✓) true or false next to each one.
Correct the false sentences.

	True	False			True	False
1. Our English class is in the morning.	☐	✓	6. We get a lot of homework.		☐	☐
2. I never come to class late.	☐	☐	7. The students sometimes eat in class.		☐	☐
3. We have class three times a week.	☐	☐	8. Our teacher drives a car to class.		☐	☐
4. There are 30 students in this room.	☐	☐	9. Cell phones often ring in class.		☐	☐
5. There is a coffee break at 10:30 a.m.	☐	☐	10. We go out for lunch after class.		☐	☐

Our English class isn't in the morning. **or** Our English class is in the evening.

2 How much do you know about your partner?

Complete the sentences to make guesses about a partner. Then ask your partner
questions to find out if you are right or wrong.

Your guesses: **Are your guesses . . .**

My partner . . .		right?	wrong?
1. _doesn't read_	a lot of books.	✓	☐
2. _____	the news on TV every night.	☐	☐
3. _____	with his / her parents.	☐	☐
4. _____	an hour a day on the Internet.	☐	☐
5. _____	at 6:00 a.m. on the weekends.	☐	☐
6. _____	tennis very well.	☐	☐

A Do you read a lot of books?
B No, I don't. But I read the newspaper every day.
A OK. I'm right about that. Do you . . . ?

3 How well do you know your city?

Complete the chart. Then use the words to tell a partner five things about your city.
Does your partner agree?

Places in a city		Words to describe places	
restaurant		expensive	

Useful expressions

a couple of	some
a lot of	no

"There are a lot of expensive restaurants in our city."   "That's right, but they're not very good."

4 Ask a question in two ways; answer more than yes or no.

A Write a second question for each question below. Start the second question with *I mean*.

1. What's your neighborhood like? _I mean, do you like it?_____

2. How often do you use a computer? _____

3. What kinds of sports do you watch on TV? _____

4. What time do you get up on weekends? _____

5. Who does the laundry at your house? _____

B *Pair work* Take turns asking and answering the questions. Say more than *yes* or *no* in your answers. Use *Well* if your answer isn't a simple *yes* or *no*.

A *What's your neighborhood like? I mean, do you like it?*
B *Well, it's not exciting, but I have nice neighbors.*

5 Are you the same or different?

A Unscramble the words to find eight kinds of TV shows.

ootrnac	cartoon	paso preoa	
mega whos		het senw	
scotmi		elarity hosw	
kalt oswh		mucrtayenod	

B Talk about your TV-watching habits with a partner. Use *Me too* and *Me neither* if you're the same. Use *Really? . . .* if you're different.

A *I never watch cartoons.*
B *Me neither. I don't like cartoons.* *Really? I love cartoons.*

6 What's your routine?

Complete each question with a verb. Can you think of four more questions? Then ask and answer with a partner.

What time do you . . .	When do you . . .
have breakfast?	_____ time with your family?
_____ to work or to class?	_____ your friends?
_____ home at night?	_____ to the movies?

How often do you . . .	Where do you . . .
_____ at the gym?	_____ your homework?
_____ a bus or train home?	_____ shopping?
_____ e-mail?	_____ lunch?

A *What time do you have breakfast?*
B *I usually eat breakfast around 7 o'clock.*

Out and about

In Unit 7, you learn how to . . .
- use the present continuous.
- talk about the weather and sports.
- ask follow-up questions to be friendly.
- use expressions like *That's great!* and *That's too bad!*

Before you begin . . .
Which of the seasons below do you have?
What's the weather usually like in the . . .

- spring?
- fall?
- rainy season?
- summer?
- winter?
- dry season?

It's hot and humid.
It's warm and sunny.
It's cool. It's often cloudy.
It's windy. It's cold.
It rains.
It snows.

San Francisco, Saturday, 3:00 p.m.: Anita is working today. Right now she's listening to her messages.

Saturday, 8:45 a.m.
Hi, Anita. This is Yoko. I'm calling from Lake Tahoe. Lisa and I are skiing today. It's snowing here. It's so beautiful! What's the weather like in San Francisco? Give me a call. Bye.

Saturday, 10:20 a.m.
Hi, it's Bill. Listen, Marcos and I are at the beach in Santa Cruz. Come and join us! Don't worry – we're not swimming. It's too cold! See you.

Saturday, 11:15 a.m.
Hey, Anita. This is Nathan. I'm in San Jose with Katie and Rob. They're playing tennis, and I'm watching. It's nice and sunny here. I hope it's not raining there. Call me! Bye.

1 Getting started

A Listen to Anita's phone messages. What's the weather like in each place?

Figure it out

B Can you complete these sentences about Anita and her friends?

1. Yoko is _____ at Lake Tahoe.
2. Marcos and Bill are at the beach, but they're _____ swimming.
3. Katie is _____ tennis with Rob, and Nathan _____ watching.
4. Anita's in San Francisco. _____ raining there.

CALIFORNIA
San Francisco
San Jose
Lake Tahoe
Santa Cruz

66

2 Grammar *Present continuous statements*

I'm	calling	from home.
You're	working	today.
She's	skiing	with a friend.
He's (not)	having	fun.
It's	raining	right now.
We're	swimming	in the ocean.
They're	playing	tennis.

The contractions isn't and aren't often follow nouns:

Marcos **isn't** working.

Marcos and Bill **aren't** swimming.

Spelling

work → work**ing**

swim → swim**ming**

have → hav**ing**

> **In conversation . . .**

In the present continuous, people usually use *'s not* and *'re not* after pronouns. People don't usually say **we aren't**, **they aren't**, **he isn't**, etc.

A Complete Anita's other phone messages.

❶ *Saturday, 12:15 p.m.* Hi, Anita. It's Joe. I hope you **'re not working** (not work) today. I _____ (not do) anything, so let's get together. Give me a call. By the way, I _____ (call) on my cell phone. See you.

❸ *Saturday, 2:50 p.m.* Hey, Anita, it's me. Chris and I are at the baseball game. It _____ (rain) right now, so they _____ (stop) the game. So, we _____ (come) over to your place. See you in 15 minutes.

❷ *Saturday, 1:00 p.m.* Hi, Anita. This is Julia. I'm at the beach with Kim. We _____ (talk) about work and things, and we _____ (have) a good time. There's no wind today, so people _____ (not windsurf). Come and join us. Bye.

B *Pair work* Student A: "Call" your partner and leave a phone message. Student B: Call back and leave a message.

"Hi, _____ . This is _____ . I'm at _____'s house. It's raining and it's cold, so we're . . ."

3 Talk about it *What's your "perfect" day?*

A Imagine you are having a perfect day. Think of answers to the questions below.

► Where are you?
► What's the weather like?
► Who are you with?
► What are you doing?

About you → **B** *Group work* Tell your classmates about your perfect day. Which day sounds like the most fun?

"I'm at the beach. It's a beautiful day! It's very hot. . . ."

67

1 Building vocabulary

A Listen to the sounds of these sports, and number the pictures. Then listen and practice.

They're playing . . .

basketball

football

volleyball [1]

They're doing . . .

aerobics

weight training

karate

They're . . .

bowling

running

biking

Word sort → **B** What sports do you play? watch on TV? Complete the chart. Compare with a partner.

I . . .	I don't . . .	I watch . . .
go bowling	play soccer	football

Notice . . .

I'm bowling right now.
I go bowling every week.

2 Building language

A Listen. Is Carl studying hard this semester? What is he doing right now? Practice the conversation.

Dad Hi, Carl. It's me. How's it going?
Carl Oh, hi, Dad. Everything's great.
Dad So, are you studying for your exams?
Carl Oh, yeah. I'm working very hard this semester.
Dad Good. So what are you doing right now? Are you studying?
Carl Uh, Dad, right now I'm watching a baseball game.
Dad Baseball? . . . Uh, who's playing?
Carl The Yankees and the Red Sox.
Dad Really? Uh, Carl, . . . let's talk again in two hours.
Carl OK, Dad. Enjoy the game!
Dad You too. But please try and study for your exams!

Figure it out → **B** Complete this question. Then ask a partner.

What sports _____ you _____ this year?

3 Grammar *Present continuous questions*

What **are** you **doing** these days?	What **is** Carl **watching** on TV?	Who**'s playing**?
Are you **studying** a lot?	**Is** he **watching** the game?	**Are** the Yankees **playing**?
Yes, I **am**.	Yes, he **is**.	Yes, they **are**.
No, I**'m not**.	No, he**'s not**.	No, they**'re not**.

A Complete the questions with the present continuous.

1. _Are_ you _getting_ (get) enough exercise these days?
2. What _____ you _____ (do) for exercise?
3. _____ you _____ (learn) a new sport?
4. How much _____ you _____ (walk)?
5. _____ you _____ (take) exercise classes this year?
6. _____ your best friend _____ (exercise) enough these days?
7. What kind of exercise _____ your best friend _____ (do)?
8. _____ your friends _____ (play) on sports teams this season?

> ···> **Time expressions**
> right now
> today
> this morning
> this week
> this month
> this year
> this semester
> this season
> these days

About you

B *Pair work* Ask and answer the questions. Give your own answers.

A *Are you getting enough exercise these days?*
B *Well, I'm playing tennis on the weekends.*

4 Speaking naturally *Stress and intonation in questions*

How often do you go to the *gym*? Are you going a *lot* these days?

A Listen and repeat the questions above. Notice how the words *gym* and *lot* are stressed. Notice how the voice falls on *gym* and rises on *lot*.

B Now listen and repeat these pairs of questions.

What's your favorite **sport**? I mean, do you like **soc**cer?
How's your favorite **team** doing? Are they doing **well** this season?
Who's your favorite **ath**lete? I mean, do you **have** a favorite?

About you

C *Class activity* Ask your classmates the pairs of questions above. What are the most popular answers?

A *What's your favorite sport? I mean, do you like soccer?*
B *No, but I like tennis.*

5 Vocabulary notebook *Who's doing what?*

See page 74 for a new way to log and learn vocabulary.

How's it going?

1 Conversation strategy *Asking follow-up questions*

A Can you add a question to this conversation?

A Hello. Are you here on vacation?
B Yes, I am. I'm here for a week.
A Really? _____ ?

🎧 Now listen. What's Kate doing this week?

Tina	Hey, Ray, this is my friend Kate. She's visiting from Chicago.
Ray	Oh, hi. Nice to meet you. So, uh . . . are you here on vacation?
Kate	Yeah. I'm here for a week.
Ray	That's great! Are you enjoying Laguna Beach?
Kate	Yeah! I'm taking a scuba-diving course.
Ray	That's cool. How's it going?
Kate	Really well. And I'm having a great time.
Tina	Oh, that's my cell phone. Excuse me.
Ray	Sure.

Notice how Ray asks Kate questions. He keeps the conversation going. Find examples in the conversation.

"I'm here for a week."

"That's great! Are you enjoying Laguna Beach?"

B 🎧 Complete the conversation with the follow-up questions. Then listen and check your answers. Practice with a partner. Can you think of more follow-up questions?

Kate So, how do you know Tina?
Ray Well, uh . . . we go to the same school.
Kate Really? _____
Ray No. I'm studying law. Actually, we play softball on the same team.
Kate Oh. _____
Ray Sure. We have ten women and six guys.
Kate That's cool. _____
Ray Every Saturday morning, when the weather's good.

Do men and women play together?

How often do you play?

Are you studying business, like Tina?

SELF-STUDY
AUDIO CD
CD-ROM

2 Strategy plus *That's . . .*

You can use expressions with *That's . . .* **to react to news.**

I'm here for a week.

That's great!

> **In conversation . . .**
>
> The top expressions for good news are:
> *That's **good** / **great** / **nice** /*
> ***interesting** / **cool** / **wonderful**.*
> The top expressions for bad news are:
> *Oh, that's **too bad** / **terrible**.*

Complete the responses using an expression with *That's . . . .* Then practice with a partner.

1. *A* I'm taking a karate class. We have a great teacher.
   B Oh, _____ .

2. *A* I'm training 8 hours a day, and I'm not sleeping.
   B Really? _____ .

3. *A* I'm playing on the college basketball team.
   B Hey, _____ .

4. *A* My friend Sarah is a professional athlete.
   B Yeah? _____ .

5. *A* Our team isn't playing well this season.
   B Oh, _____ .

6. *A* I'm reading a book about the history of the World Cup.
   B Really? _____ .

3 Listening and speaking *How's your week going?*

A 💿 Listen. Six people tell you about their week. Respond to each person using an expression with *That's . . . .*

1. _____   4. _____
2. _____   5. _____
3. _____   6. _____

About you ➤ **B** *Pair work* Student A: What are you doing these days? Tell your partner. Student B: Listen and respond with *That's . . . .* Ask two follow-up questions. Then change roles.

A I'm taking a swimming class.
B That's nice. Do you like your teacher?

4 Free talk *What's hot? What's not?*

See *Free talk 7* at the back of the book for more speaking practice.

Staying in shape

1 Reading

A Which of these statements are true for you? Tell the class.

	True	False
I walk to school / to work every day.	☐	☐
I walk around my neighborhood a lot.	☐	☐
I think walking is boring.	☐	☐

	True	False
I never walk in the rain.	☐	☐
I use the stairs, not the elevator.	☐	☐
I go hiking on the weekend.	☐	☐

B Read the article. Why does the author think walking is a good idea?

DON'T WAIT – JUST WALK!

So you're not getting enough exercise? And you hate sports, and you can't stand the gym?

Well, if you're looking for a new exercise routine, try walking. Here are six reasons why walking is a great idea.

1 Walking is easy.

You just walk – left, right, left, right. See? It's easy.

2 Walking is cheap.
Don't spend money on expensive clothes and equipment. All you need is a pair of good shoes or sneakers.

3 Walking gives you time for yourself.

Listen to a CD or a book, think about life, relax.

4 Walking is good for you.

You feel good after a long walk. And now science is proving that walking outdoors is the best exercise.

5 Walking is fun.

Go with a friend. Walk and talk! What's going on in your neighborhood? Walk around and find out!

6 Walking is good in all kinds of weather.
So it's raining? Don't worry, take an umbrella. When it's sunny, use sunscreen. If it's hot, take some water with you. When it's cold, you always feel warm.

C Add these missing sentences to the paragraphs above.

✓ Walk and talk!

It's good for your mind and body.

And there's no gym membership fee.

There are no special instructions, and there are no rules.

And walking in the snow is great exercise!

Do something you enjoy.

D Read the article again. Do you agree that walking is good exercise? What are the three best reasons the author gives?

2 *Listening* *Do you enjoy it?*

A Listen to the conversations. Number the pictures from 1 to 4.

1

B Listen again. Why do the people enjoy their exercise? Write one reason below each picture.

3 *Writing* *An article for a health magazine*

A Think of an exercise you enjoy. Write a short article like the one on page 72.
Think of a title and three headings. Write at least two sentences for each heading.

Try Aerobics!

1 **Aerobics is fun.**
Find a fun teacher. Talk to your classmates, and make new friends. Don't be shy! . . .

2 **Aerobics is good for you.**
It's good for your heart. And you feel good after class. . . .

3 **Aerobics is easy.**
Buy an aerobics video, and exercise in front of the TV. Do it before breakfast. . . .

> **Help note**
>
> *Imperatives for advice*
>
> *Find* a fun teacher.
> *Make* new friends.
> *Don't* be shy!

B *Group work* Read your classmates' articles. Which type of exercise sounds like fun?

Vocabulary notebook

Learning tip *Writing true sentences*

To remember new vocabulary, use words in true sentences.

1 Complete the sentences about the weather.

1. Outside right now, it _____ .
2. At this time of year, it usually _____ .
3. In the summer, it _____ .
4. In the winter, it _____ .
5. I like the weather when it _____ ,
   but I don't like it when it _____ .

> **It's cold outside!**
>
> In the U.S. and Canada, the top 6 weather expressions with ***it's*** are:
> 1. It's cold.
> 2. It's hot.
> 3. It's raining.
> 4. It's windy.
> 5. It's humid.
> 6. It's snowing.
>
> People say ***It's cold*** 10 times more than ***It's hot***.

2 Write the names of six people you know. Complete the chart with true sentences.

Name	Where is he or she right now?	What is he or she doing right now?	What sports or exercise is he or she doing these days?
my brother Ming	He's at school.	He's studying math right now.	He's playing soccer and basketball.
➊			
➋			
➌			
➍			
➎			
➏			

On your own

Take a minute this week, and look around you. What are people doing? Write 6 sentences.

He's snoring

74

Shopping

In Unit 8, you learn how to . . .

- use *like to, want to, need to,* and *have to* with other verbs.
- use *this, that, these,* and *those*.
- ask questions with *How much*.
- talk about clothes, colors, shopping, and prices.
- use expressions like *Um* . . . and *Let's see* . . . .
- use "conversation sounds" like *Uh-huh* and *Oh*.

Before you begin . . .

Look at the pictures. Find the people who are wearing these clothes.

- pants and a top
- jeans and a T-shirt
- a suit and tie
- a dress and high heels

Are any of your classmates wearing these things today?

What kinds of clothes do you like to wear?

**Kayo Noguchi, 16,
high school student**

**Rick Govia, 27,
accountant**

**Louisa Vandermeer, 32,
advertising executive**

Well, we don't have to wear
uniforms at our school, so I like
to wear pants, a T-shirt, and
sneakers.

I have to wear a suit and tie to
work. After work, I just want
to go home and put on jeans
and an old sweater. You know,
something comfortable.

I like to wear designer clothes,
because I need to look good for
work. So I usually wear dressy
pants or a nice skirt, with a jacket
and a silk blouse. Oh, and of
course, high heels.

1 Getting started

A Listen. What kinds of clothes do Kayo, Rick, and Louisa usually wear?

**Figure
it out →**

B Circle the words to make true sentences about the people above. Then make three
true sentences about yourself. Tell a partner.

1. Kayo **has to / doesn't have to** wear a uniform.
2. Rick **wants to / doesn't want to** wear comfortable clothes after work.
3. Louisa **needs to / doesn't need to** wear nice clothes to work.

2 Grammar *Like to, want to, need to, have to*

> What kinds of clothes **do** you **like to** wear?
> I **like to** wear casual clothes.
> I **don't like to** wear suits.
>
> **Do** you **have to** wear a suit to work?
> Yes, we do. We **have to** wear a suit and tie.
> No, we don't. We **don't have to** wear a suit.
>
> What **do** you **want to** wear tonight?
> I **want to** wear my new outfit.
> I **don't want to** wear my old dress.
>
> **Do** you **need to** buy new shoes?
> Yes, I do. I **need to** get some sneakers.
> No, I don't. I **don't need to** buy new shoes.

About you →

A Make true sentences. Then compare with a partner.

1. I _don't like to_ (like to) wear jeans all the time.
2. Our teacher _____ (have to) wear a suit to class.
3. The students in this class _____ (have to) wear a uniform every day.
4. I _____ (want to) buy shoes this weekend.
5. My parents _____ (like to) spend a lot of money on clothes.
6. I _____ (need to) buy new jeans.
7. My best friend _____ (like to) shop for clothes.
8. We _____ (need to) wear warm clothes in the spring.

"I don't like to wear jeans all the time."   *"Me neither. But I like to wear jeans on the weekends."*

B Write five questions with *like to*, *want to*, *need to*, and *have to*.
Then ask a partner your questions.

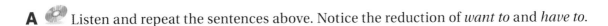

3 Speaking naturally *Want to and have to*

> /wanə/
> I **want to** buy some new clothes.
> What do you **want to** buy?
>
> /hæftə/
> I **have to** buy some new clothes.
> What do you **have to** buy?

A Listen and repeat the sentences above. Notice the reduction of *want to* and *have to*.

B Now listen and repeat these questions.

1. Do you have to go shopping this week? . . . Where do you have to go?
2. Do you have to buy any new clothes? . . . What do you have to get?
3. Do you want to spend a lot of money? . . . How much do you want to spend?
4. Do you want to go to a designer store? . . . Which stores do you want to go to?

About you →

C *Pair work* Ask and answer the questions. What interesting things do you find out?

A **Do you have to go shopping this week?**
B **Yes, I have to go on Saturday.**
A **Where do you have to go?**

Things to buy

1 Building vocabulary

A Listen and say the words. Which of these items do you have? Which do you need to buy? Tell the class.

a baseball cap

a belt

a backpack

a briefcase

a purse

shoes and socks

a bracelet and a ring

a necklace and earrings

a coat and boots

a hat, a scarf, and gloves

Word sort

B What clothes and accessories do you have in these colors? Write them in the chart. What colors do you like to wear? Discuss with a partner.

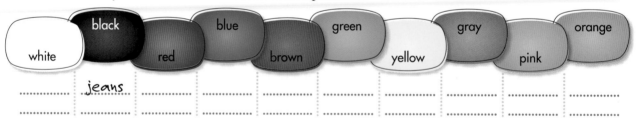

white	black	red	blue	brown	green	yellow	gray	pink	orange
	jeans								

"I like to wear black. I have some black jeans and a black jacket."

2 Building language

A Listen. How much are the gloves and the scarf? Practice the conversation.

Salesperson Hello. Can I help you?

Samantha Uh, hi. How much are those gloves?

Salesperson These? They're really popular. They're $80.

Samantha Hmm. And what about that blue scarf? How much is that?

Salesperson This scarf is on sale. It's only $149.

Samantha $149? OK, I have to think about it. Thanks anyway.

Figure it out

B *Pair work* Can you act out the conversation using your own ideas?

3 Grammar *How much . . . ?; this, these; that, those*

How much is **this** (scarf)? ·········⟶ are **these** (gloves)? **How much** is **that** (watch)? ·········⟶ are **those** (sunglasses)? **How much** is **it**? **It's** $2.99. are **they**? **They're** $49.99.	***Saying prices*** $2.99 = "Two dollars and ninety-nine cents" or "Two ninety-nine" $125 = "A hundred and twenty-five (dollars)" $475 = "Four hundred and seventy-five (dollars)"

In conversation . . .

People also say *How much does it cost?* and
How much do they cost? to talk about prices in general.

A Write questions with *How much . . . ?* and *this, that, these,* and *those.* Then ask and
answer the questions with a partner.

$239.00

$ 14.99

$29.98

1. How much are
   these sunglasses ?

2. _____
   _____ ?

3. _____
   _____ ?

GIANTS
$10.50

$119.00

$49.98

4. _____
   _____ ?

5. _____
   _____ ?

6. _____
   _____ ?

B *Group work* Talk about how much these things cost. Agree on an average price.

  shoes   a cell phone   jeans   a CD   a backpack

"How much do shoes cost?"   *"Well, it depends. They're about thirty or thirty-five dollars."*

4 Vocabulary notebook *Nice outfit!*

See page 84 for a new way to log and learn vocabulary.

Can I help you?

1 Conversation strategy Taking time to think

A When you need time to think, what do you say in your language?
Do you have expressions like *Um*, *Well*, and *Uh*?

> A *How often do you go shopping?*
> B *Um, well. Uh . . . once or twice a month maybe.*

Now listen. What does Sarah buy? How much is it?

Clerk	Can I help you?
Sarah	Uh, yes. I'm looking for a bracelet.
Clerk	All right. Is it a gift?
Sarah	Uh-huh, it's a birthday present for a friend.
Clerk	OK. And how much do you want to spend?
Sarah	Well, let's see . . . about $40, I guess.
Clerk	Uh-huh. Well, we have these silver bracelets here.
Sarah	Oh, this looks nice. Um . . . how much is it?
Clerk	Um, let's see . . . it's $55.95.
Sarah	Oh. That's a lot. Let me think . . . OK. I guess I'll take it.

Notice how Sarah says *Uh, Um, Well, Let's see*, and *Let me think* when she needs time to think. Find examples in the conversation.

"How much do you want to spend?"

"Well, let's see . . ."

B *Pair work* Student A: You are a salesperson. Help a customer. Student B: You are a customer. You are shopping for one of the items below. Have a conversation.

a watch   a camera   a backpack   a pair of sunglasses

A *Can I help you?*
B *Um, yeah. I'm looking for a watch.*
A *OK. And how much do you want to spend?*
B *Well, let me think . . . about $100.*

SELF-STUDY
AUDIO CD
CD-ROM

2 *Strategy plus* *"Conversation sounds"*

Uh-huh means "Yes," "That's right," or "I'm listening."

Oh shows you're surprised, happy, or angry.

This bracelet is $55.95.

Is it a gift?

Uh-huh.

In conversation . . .

Uh-huh and **Oh** are in the top 50 words.

Oh. That's a lot.

A Look at these conversations. What do the "conversation sounds" mean?

1. *A* You have some money with you, right?
   B Let's see . . . I have about sixty dollars.
   A **Oh**, good! Can I borrow ten dollars?
   B **Oh**, not again!

2. *A* How much do you usually spend on a sweater? About eighty dollars?
   B **Uh-huh**. Seventy or eighty dollars.

3. *A* I have to go shopping this weekend.
   B **Uh-huh**.
   A I want to buy a flat-screen TV.
   B **Oh**, cool!

4. *A* How many credit cards do you have?
   B Nine or ten.
   A **Oh**, that's a lot!
   B **Uh-huh**. I know, but I never carry cash.

About you

B *Pair work* Practice the conversations. Give your own reactions and answers.

3 *Listening* *I'll take it.*

A Listen to three conversations in a store. Write the prices of each item.

B Listen again. Which items do the shoppers buy? Circle the items.

4 *Free talk* *How do you like to dress?*

See *Free talk 8* at the back of the book for more speaking practice.

Shop till you drop!

1 Reading

A Brainstorm words related to shopping. How many words can you think of?
Remember, there are no right or wrong answers!

clothes ——————— **shopping** ——— money

B Read the article. Can you find any of the words you brainstormed?

Shopping around the world

How do you like to shop? In a mall with over 800 stores? Or in a traditional market? If you're planning a trip to any of these countries, read about these great places to shop.

Italy

San Lorenzo Market in Florence is famous for leather purses, wallets, and gloves. There are lots of cafés – so after a morning of shopping, enjoy a coffee in the historic city center.

Japan

You need all day to shop at *Takashimaya Times Square, located in the Shinjuku section of Tokyo.* At this department store, there are boutiques with designer clothing, gifts, and housewares. There's also an art gallery and a travel agency. And best of all, there are three floors of restaurants!

Canada

The West Edmonton Mall in the province of Alberta is the size of 48 city blocks and has over 800 stores. And everything is cheap, because you don't have to pay sales tax. There are also 100 restaurants, 26 movie theaters, amusement parks, nightclubs, a hotel, and even an ice-skating rink!

Morocco

In the city of Marrakech, there's the famous "souk." There are hundreds of stalls selling clothes, traditional Moroccan slippers, and copper

pots. Choose gifts for all of your friends. But remember – when you buy something, you have to bargain, because there are no prices!

C Look at the article again. Find . . .

▥ a good place to have coffee.

▥ a place with no sales tax.

▥ a place where you have to agree on the price.

▥ a place where you can shop and then look at art.

▥ somewhere you want to go.

▥ something you want to buy.

▥ four words for a place to shop.

▥ four new words you want to learn.

2 Listening and writing *Favorite places to shop*

A What's your favorite store? Why do you shop there? Tell the class.

B 💿 Listen to Min Sup talk about his favorite store. Circle the correct information.

1. I like to shop **at the mall** / **in small stores**.
2. My favorite store is **a bookstore** / **an electronics store**.
3. I like it because it's **cheap** / **interesting**.
4. I usually go there **on Friday nights** / **on Sundays**.
5. I buy a lot of **DVDs** / **computer games**.

C Write a paragraph about your favorite store for a shopper's guide. Use the model below.

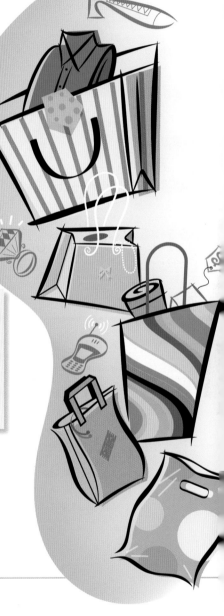

My favorite store is . . .
I like it because . . .
I usually go there . . .
They sell . . .
They also have . . .
I like to buy . . .

> **Help note**
>
> **Linking ideas with** *because* **to give reasons**
>
> *My favorite store is Chang's, **because** it has great clothes. I like to shop there **because** it's cheap.*

D *Group work* Take turns reading your recommendations. Do you learn about any new places to shop?

3 Talk about it *What kind of shopper are you?*

Group work Discuss the questions. Does anyone have an interesting shopping habit?

▶ Do you like to go shopping? Why or why not?

▶ Do you usually pay cash or with a credit card?

▶ Do you compare prices before you buy something?

▶ Do you like to buy things on sale?

▶ Do you ever buy things you don't need?

▶ Do you ever spend too much money?

Nice outfit!

Learning tip *Labeling pictures*

To learn new vocabulary, you can label pictures in books or magazines.

1 Label the clothing and personal items in this picture.

necklace

2 Find and label at least three pictures you like from a magazine or catalog.

On your own

Go into a big clothing store. How many things can you name in English?

A wide world

In Unit 9, you learn how to . . .
- use *can* in statements, questions, and short answers.
- talk about things you can do in your city.
- talk about countries, languages, and nationalities.
- talk about international foods.
- explain the meaning of a word.
- use *like* in different ways.

Pelourinho – The Historic District of Salvador, Brazil

The Sydney Harbour Bridge in Australia

The Eiffel Tower in Paris, France

The Grand Palace in Bangkok, Thailand

The Pyramids of Chichen Itzá, Mexico

Before you begin . . .
Do you enjoy sightseeing? Check (✓) the activities you like to do.

- see famous bridges or statues
- visit interesting or historic areas
- go to the tops of towers
- go to palaces or castles

Five Great Things to Do in **New York**

Take a ferry to the Statue of Liberty, and buy souvenirs. ▶

Go to a Broadway show, and see Times Square. ▶

◀ Visit one of over 50 museums.

▲ Take a walk through Central Park.

◀ Get a view of the city from the top of the Empire State Building.

Emma Oh, no. It's raining! What can you do in New York on a day like this?

Ethan Oh, come on. You can do a million things. We can take a ferry to the Statue of Liberty.

Emma A ferry – in this weather?

Ethan Well, . . . we can go to the top of the Empire State Building.

Emma But you can't see anything in the rain.

Ethan Yeah, you're right. I know – let's go to a Broadway show. There are shows on Wednesday afternoons.

Emma OK. It's a deal. But first can we get an umbrella?

1 Getting started

A Look at the page above from a guidebook. Choose two fun things to do. Tell the class.

B 💿 Listen. What do Ethan and Emma decide to do? Practice the conversation.

Figure it out → **C** *Pair work* What are some things you can do in New York City? Take turns giving ideas.

"You can see the Statue of Liberty."

2 Grammar *Can and can't*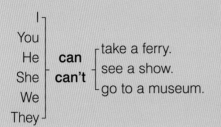

I		take a ferry.
You		
He	**can**	see a show.
She	**can't**	
We		go to a museum.
They		

What **can** you do in New York?
You **can** do a million things.

Can we buy an umbrella?
Yes, we **can**.
No, we **can't**.

> **In conversation . . .**
>
> **You** is the most common word before **can**. **You** often means "people in general."
>
> **You** *can't take pictures.* = *It's not possible to take pictures.*

A What can you do in New York City? Match the questions and answers. Then practice with a partner.

1. What historic neighborhoods can you see? __b__
2. Can you take a walking tour? _____
3. What historic sites can you visit? I mean, can you go to a castle? _____
4. What can you do on a rainy day? _____
5. What kinds of museums can you go to? _____
6. Where can you get a good view of the city? I mean, can you go to the top of a tall building? _____

a. Yes, you can. You can go to the top of the Empire State Building.
b. You can walk around Greenwich Village. It's a beautiful neighborhood.
c. Yes, and you can take a bus tour, too.
d. Well, you can go to an art museum or to the Museum of Natural History.
e. You can go shopping or go to a museum.
f. No, you can't. There are no real castles in New York.

> **About you** →

B *Pair work* Ask the questions again, and give answers about your own city.

A **What historic neighborhoods can you see in Mexico City?**
B **Let me think . . . you can go to the Zona Rosa – the Pink Zone.**

3 Speaking naturally *Can and can't*

/kən/	/kən/	/kæn (t)/
*What **can** you do here?*	*You **can** go to the zoo.*	*You **can't** go on Mondays.*

A  Listen and repeat the sentences above. Notice the pronunciation of *can* and *can't*.

B Listen and complete the sentences with *can* or *can't*.
Then make them true for your city. Discuss with a partner.

1. You _____ see a lot of famous people.
2. You _____ spend a day at the beach.
3. You _____ take a ferry to an island.
4. You _____ sit at outdoor cafés at night.
5. You _____ see a different movie every night.
6. You _____ go to a show or concert every weekend.

A **You can't see a lot of famous people in our city.**
B **No, you can't.**

1 Building vocabulary

A Listen and say the countries and regions. Which do you know in English? Check (✓) the boxes. What other countries do you know?

■ North America
■ Canada
☐ the United States
☐ Mexico
■ Central America
☐ Costa Rica
☐ Peru
■ South America
☐ Chile
■ Europe
☐ Germany
☐ Great Britain
☐ France
☐ Spain
☐ Italy
■ The Caribbean
☐ Puerto Rico
☐ Morocco
☐ Brazil
■ Africa
■ Asia
☐ Russia
☐ South Korea
☐ Japan
☐ China
☐ Turkey
☐ Thailand
☐ India
☐ Kenya
☐ Australia
☐ Oceania
☐ New Zealand
☐ Antarctica

B Where do people speak these languages? Use ideas from the map. Then compare your answers with a partner.

| English | Spanish | Turkish | Chinese | Portuguese | Japanese |
| Italian | Korean | Russian | German | Arabic | French | Thai |

"They speak English in the United States, Canada, . . ."

C Complete the chart with languages and countries. Compare your answers with a partner.

I can speak . . .	I can't speak . . .	I want to go to . . .
Portuguese a little English	Korean	Australia

Can for ability

*I'm Brazilian. I speak Portuguese, and I **can** speak a little English, but I **can't** speak Korean.*

"I can speak a little English, so I want to go to Australia."

88

2 Listening and speaking *National dishes*

A Do you like food from other countries? What kinds of food do you like? Tell the class.

B Listen to Melissa talk about food. What types of food does she like? Check (✓) the boxes.

☐ French   ☐ Korean   ☐ Italian   ☐ Chinese

☐ Japanese   ☐ Brazilian   ☐ Mexican   ☐ Thai

About you → **C** *Pair work* Ask and answer questions about international foods. Add your own ideas.

"*Do you like French food?*"

"*Can you make Korean food?*"

"*Where can you get good Italian food around here?*"

Countries	Nationalities
Japan	Japanese
Spain	Spanish
Mexico	Mexican
Peru	Peruvian

3 Survey

Do your classmates know a lot about other countries and cultures? Ask questions to find out.

Who is a "world citizen"?

Find someone who . . .	Name
can make Indian food.	_____
has an American friend.	_____
can speak three languages. (What are they?)	_____
likes Brazilian music.	_____
knows the capital city of Australia.	_____
can name three countries beginning with **C**.	_____
knows the name of a British band.	_____

"*Can you make Indian food?*"   "*Yes, I can make some things.*"   "*No, I can't. I never cook.*"

4 Vocabulary notebook *People and nations*

See page 94 for a new way to log and learn vocabulary.

Lesson C — They're a kind of candy.

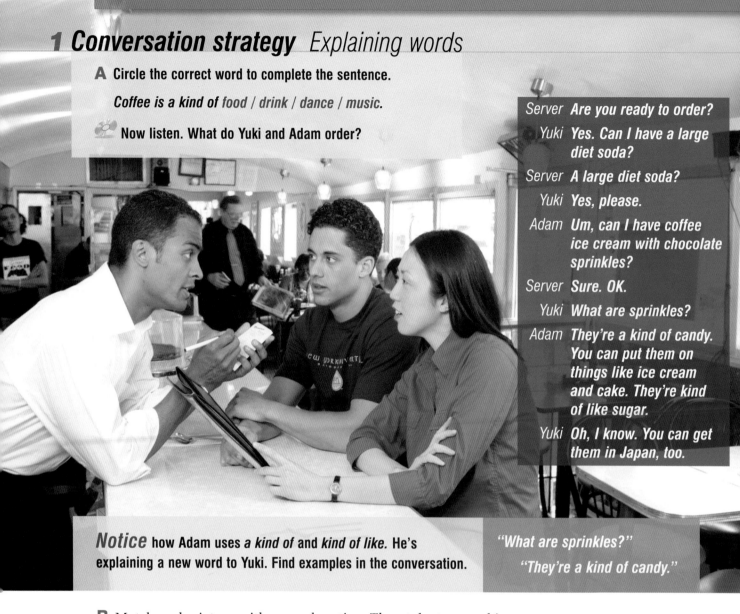

1 Conversation strategy Explaining words

A Circle the correct word to complete the sentence.

Coffee is a kind of food / drink / dance / music.

Now listen. What do Yuki and Adam order?

Server	Are you ready to order?
Yuki	Yes. Can I have a large diet soda?
Server	A large diet soda?
Yuki	Yes, please.
Adam	Um, can I have coffee ice cream with chocolate sprinkles?
Server	Sure. OK.
Yuki	What are sprinkles?
Adam	They're a kind of candy. You can put them on things like ice cream and cake. They're kind of like sugar.
Yuki	Oh, I know. You can get them in Japan, too.

Notice how Adam uses *a kind of* and *kind of like*. He's explaining a new word to Yuki. Find examples in the conversation.

"What are sprinkles?"

"They're a kind of candy."

B Match each picture with an explanation. Then take turns asking a partner to explain the words.

1 *lassi* (from India)
2 a *bouzouki* (from Greece)
3 a *hanbok* (from Korea)
4 a *tortilla* (from Mexico)

- [ ] It's a kind of musical instrument. It's like a guitar.
- [ ] It's a kind of bread. It's kind of like a pancake.
- [ ] It's a kind of drink. It's kind of like a milk shake.
- [ ] It's a kind of dress. It's a traditional outfit.

"What's lassi?"   "It's a kind of . . ."

SELF-STUDY AUDIO CD CD-ROM

90

2 *Strategy plus* *Like*

You can use *like* to give examples.

You can put sprinkles on things like ice cream and cake.

> **In conversation . . .**
>
> *Like* is one of the top 15 words.
> It has other meanings:
>   *I **like** Brazilian food.*
>   *What's Thailand **like**?*
>   *Sprinkles are **like** sugar.*

Imagine a tourist is asking these questions about your country.
Answer the questions. Then practice with a partner.

1. *A* What are good souvenirs to buy?
   B Well, things like _____ .

2. *A* Do you ever see people in traditional clothes?
   B Well, you sometimes see things like _____ .

3. *A* Is the food spicy like Indian food?
   B _____ .

4. *A* Where are good places to visit?
   B Oh, places like _____ .

3 *Listening and speaking* *What language is it from?*

A Label the photos with the words below. Can you guess what languages
the words are from? Complete the chart.

	❶	❷	❸	❹
a mosquito				
✓ yogurt				
a boutique				
a waltz				
It's called . . .	yogurt			
The word is from . . .				

B Now listen to more of Adam and Yuki's conversation. Check your guesses.

C *Group work* What foreign words do you use in your language? How many
can you think of? What do they mean? Make a list.

4 *Free talk* *Where in the world . . . ?*

See *Free talk 9* for more speaking practice.

Exciting destinations

1 Reading

A What do you know about these popular tourist destinations? What can you do or see there? Make a class list.

London   Tokyo   San Francisco   Rome

B Look at the Web page. How many of your ideas are mentioned?

The Travel Guide

  the travel guide

Search for more great destinations.

SEARCH [   ] GO

Go on a Virtual Tour! Are you planning a trip to one of the popular destinations below? Check out our guide to these great cities. You know the famous places, but with the Travel Guide, you can see and do something unusual. Click on one of the links below.

London
London is famous for historic buildings like Buckingham Palace. But you can also take a walking tour of its haunted buildings. The Tower of London is just one example. Can ghosts speak English? Go and find out. More . . .

Tokyo
Everyone loves to eat sushi. But where does all that fish come from? Go to the Tsukiji Fish Market at 5:00 a.m., and find out. It's lively, colorful, and fun. And you can have great sushi for breakfast! Then go to the Ginza. More . . .

San Francisco
See the Golden Gate Bridge, and then walk down to Golden Gate Park. Go to the Japanese Tea Garden, and enjoy the beautiful waterfalls, bonsai trees, and plants. Then order tea from a waitress in a kimono. More . . .

Rome
After your tour of the Colosseum and St. Peter's Basilica, hang out on the Spanish Steps with young people from around the world. More . . .

Travel Competition
Win a trip to Hawaii.
Click here now!

Do you have an unusual photo of a popular tourist attraction? Enter our **photo competition**. Win a digital camera.
Click here for details.

For exchange rates, click here.

Click here for world weather.

Do you have a good travel tip? Share your stories on our message board.

Need some help?
Our trip expert can help you find your ideal vacation.
Click here.

C Look at the Web page again. Find these things and answer the questions. Then compare with a partner.

Find . . .

▥ the links for two competitions. What can you win?

▥ two other kinds of information. Why is the information useful?

▥ the place to send your own travel tip. Do you have a good travel tip?

▥ two places you want to read more about. What do you want to know?

2 *Talk about it* *Do you want to take a trip?*

Group work What ideas do you and your classmates have about travel?

Can you agree on . . .

▶ three countries you all want to go to?

▶ the three best places to visit in your country?

▶ three tourist attractions you want to see?

▶ three types of food you all want to try?

▶ two languages you need when you travel abroad?

▶ three really good souvenirs to buy?

A Well, I want to go to Chile.

B Oh, that's a good idea.
We can go to the Atacama Desert.

C Yeah, but I can't speak Spanish.
What about . . . ?

The Atacama Desert

3 *Writing* *A paragraph for a Web page*

A Write about a place you know well for the Travel Guide Web page.

Bangkok
Bangkok is famous for its palaces, temples, and beautiful river. First go to the Grand Palace. Then visit the many historic temples. After you tour the city, you can take a boat trip on the river. You can enjoy the sunset and then see the temples by night.

> **Help note**

Commas in lists

It's lively, colorful, and fun.

Enjoy the beautiful waterfalls, bonsai trees, and plants.

B *Class activity* Read your classmates' paragraphs. Which ones are the most interesting?

People and nations

Learning tip *Grouping vocabulary*

You can sort new vocabulary into different types of groups. You can group nationalities by their endings and countries by their regions.

1 Choose 15 or more nationalities you want to learn. Write them in the chart.

-ese	-ian / -an / -n
Vietnamese	Colombian

-ish	other
Spanish	Greek

2 Complete the chart with different countries. How many countries can you think of in each region?

Africa	Asia	Europe
Morocco	Thailand	France

North America	Central America	South America

Some countries and nationalities

Argentina	Argentine
Australia	Australian
Brazil	Brazilian
Canada	Canadian
Chile	Chilean
China	Chinese
Colombia	Colombian
Costa Rica	Costa Rican
Ecuador	Ecuadorian
Egypt	Egyptian
France	French
Germany	German
Great Britain	British
Greece	Greek
Guatemala	Guatemalan
Iraq	Iraqi
Ireland	Irish
Israel	Israeli
Italy	Italian
Japan	Japanese
Lebanon	Lebanese
Morocco	Moroccan
Panama	Panamanian
Peru	Peruvian
Poland	Polish
Portugal	Portuguese
Russia	Russian
South Korea	South Korean
Spain	Spanish
Thailand	Thai
Turkey	Turkish
Venezuela	Venezuelan
Vietnam	Vietnamese

On your own

Find a world map. Label it in English. How many countries do you know?

1 Questions and follow-up questions!

A Complete the questions with verbs. Then match the questions and answers. Practice with a partner.

1. What __are__ you __wearing__ today? (wear) __d__
2. What colors _____ the teacher _____ today? (wear) ____
3. What _____ in your neighborhood this week? (happen) ____
4. What can you _____ in your neighborhood after midnight? (do) Can you _____ dancing? (go) ____
5. What do you want _____ tonight? (do) ____
6. What kinds of restaurants do you like _____ to? (go) ____
7. What languages can you _____? (speak) ____
8. What do you have _____ next weekend? (do) ____
9. What time do you have _____ tomorrow? (get up) ____
10. What _____ your friends _____ today? (do) ____
11. How often do you like _____ your family? (see) ____
12. What _____ you _____ about right now? (think) ____

a. There's a concert.
b. I want to stay home.
c. Every weekend.
d. Jeans and a sweater.
e. Food. I'm hungry.
f. Blue and gray.
g. English and a little Spanish.
h. They're all working.
i. I need to clean the house.
j. Well, I like Thai and Italian food.
k. No, you can't, but you can see a movie.
l. Early. I have to be at work before 8:00.

B *Pair work* Choose five questions and start conversations. Ask follow-up questions. How many follow-up questions can you ask for each topic?

A *What languages can you speak?*
B *I can speak Spanish and English.*
A *Can you speak Portuguese?*
B *No, I can't, but I can read and understand it.*

2 Play a word game.

Complete the chart. Write a word for each category beginning with each letter. You have two minutes! Then compare with a partner. Who has a word in every space?

Category	S	B	T	R
a sport	soccer			
a country				Russia
a nationality		Brazilian		
an item of clothing or jewelry			tie	

A *What sport begins with "S"?*
B *Let's see. I have "skiing."*
A *I have "soccer."*
B *OK, what country begins with "S"?*

3 *Can you use these expressions?*

Complete the conversation. Use the expressions in the box. Then practice with a partner.

this	those	kind of like	Let me think	✓ That's great
that	like	a kind of	Let's see	That's too bad

Samir Grant! What are you doing here?

Grant I'm working here for the summer.

Samir Wow! __That's great__ . Hey, I like your uniform.
I mean, _____ shirt is cool.

Grant Yeah, but I can't stand _____ hat. It's so hot.

Samir _____ . Do you have to wear it?

Grant Uh-huh. So, what can I get for you?

Samir _____ . . . what do you have?

Grant Um . . . we have things _____ ice cream,
frozen yogurt, smoothies, . . .

Samir What's a smoothie?

Grant It's _____ drink.
It's _____ a milk shake.

Samir _____ . Do I want frozen yogurt
or a smoothie?

Grant Well, they're both good.

Samir Hey, do people really buy _____ hats?

Grant Actually, they're free with the frozen yogurt.

Samir In that case, can I have a smoothie?

4 *Do you have similar interests and tastes?*

A Complete the sentences in the chart with your own information.

Sports	**Countries and languages**
I don't like to watch _____ .	*I want to go to* _____ .
I want to learn (to) _____ .	*I want to learn* _____ .
Colors	**Clothes**
I like to wear _____ .	*I never wear* _____ .
I can't wear _____ .	*I wear* _____ *a lot.*
Seasons	**Weather**
I love the _____ .	*I hate to go out when it* _____ .
I don't like the _____ .	*I love to be outside when it* _____ .

B **Group work** Compare sentences. What do you have in common?

A I don't like to watch golf on TV.

B Me neither. I think it's boring.

C Really? I love to watch golf. But I don't like to watch baseball.

Self-check

How sure are you about these areas?
Circle the percentages.
grammar
20% 40% 60% 80% 100%
vocabulary
20% 40% 60% 80% 100%
conversation strategies
20% 40% 60% 80% 100%

. . . .

Study plan

What do you want to review?
Circle the lessons.
grammar
7A 7B 8A 8B 9A 9B
vocabulary
7A 7B 8A 8B 9A 9B
conversation strategies
7C 8C 9C

Busy lives

In Unit 10, you learn how to . . .

- use the simple past of regular and irregular verbs.
- talk about things you did last night and last week.
- use expressions like *Congratulations* and *Good luck*.
- say *You did?* to show that you're interested or surprised.

Before you begin . . .

Are you usually busy in the evening?

Do you do any of these things?

What else do you do?

A night at home

What did you do last night?

Josh Let me think. I stayed home, played a video game, and listened to a new CD. That's it.

Mari I tried to study for a math exam while my roommate practiced her flute.

Peter Well, my wife rented a DVD, so we watched a movie. But I didn't like it much.

Melissa I didn't want to go out, so I invited a couple of friends over, and we cooked dinner.

Rachel Oh, I just worked late and then cleaned the house. You know – the usual.

Stephen I chatted online with my friend Jay. He's living in Italy.

1 Getting started

A Listen and read. Who had fun last night? Who didn't?

Figure it out → **B** Can you complete these sentences about the people above?

1. Josh _____ to music last night.
2. Mari _____ for an exam.
3. Peter and his wife _____ a movie.
4. Melissa _____ some friends over for dinner.
5. Rachel _____ late.
6. Stephen and Jay _____ online.

2 Grammar *Simple past statements – regular verbs*

I	**played**	a video game.	I	**didn't play**	chess.
You	**studied**	math.	You	**didn't study**	English.
He	**watched**	a movie.	He	**didn't watch**	TV.
She	**wanted**	to stay home.	She	**didn't want**	to go out.
We	**cooked**	Italian food.	We	**didn't cook**	Chinese food.
They	**chatted**	online.	They	**didn't chat**	very long.

Past tense endings

watch → watch**ed**
invite → invite**d**
play → play**ed**
study → stud**ied**
chat → chat**ted**

About you → What did you do last night? How about your friends and family?
Make true sentences. Then compare with a partner.

1. I __didn't watch__ (watch) TV last night.
2. I _____ (clean) the house.
3. My friends and I _____ (chat) online.
4. I _____ (study) English.
5. I _____ (play) a computer game.
6. My best friend _____ (invite) me over for dinner.
7. My neighbor _____ (call) me.
8. My family _____ (stay) home.

A I didn't watch TV last night. How about you?
B Well, let's see . . . I watched the news.

In conversation . . .

People use the simple
present and simple past more
often than any other tense.

3 Speaking naturally *-ed endings*

/t/ *I worked on Saturday.*   /d/ *We played a game.*   /ɪd/ *I chatted online.*

A Listen and repeat the sentences above. Notice the *-ed* endings of the verbs.

B Listen to these sentences. Do the verbs end in /t/, /d/, or /ɪd/?
Check (✓) the correct column.

Last night . . .	/t/	/d/	/ɪd/
1. I cooked a big meal.	✓	☐	☐
2. I rented a DVD.	☐	☐	☐
3. I played a video game.	☐	☐	☐
4. I watched a movie.	☐	☐	☐
5. I e-mailed a couple of friends.	☐	☐	☐

About you → **C Group work** Make five true sentences about last night.
Tell your classmates. What are the most popular evening activities?

"Last night I played basketball. . . ."

A busy week

1 Building vocabulary *Irregular verbs*

A 💿 Listen and say the sentences. Did you do these things last week? Tell the class.

"I didn't buy a sweater, but I bought some CDs."

❶ I **bought** a sweater.

❷ I **had** a piano lesson.

❸ I **made** a lot of phone calls.

To call:
~~Richard K.~~
~~Suzie Q.~~
~~rces B.~~

❹ I **saw** three movies.

❺ I **read** a couple of books.

❻ I **went** to a party.

❼ I **took** an exam.

FINAL EXAM

❽ I **met** someone interesting.

❾ I **did** a lot of work.
I **wrote** three reports.

REPORT

Word sort → **B** Write three things you did at each time below. Then compare with a partner.

last Sunday	last Friday	last week
I saw a movie.		

"Last Sunday I saw a movie."   *"Really? I went to a party."*

2 *Building language*

A  Listen to the questions. Check (✓) the answers that are true for you.

Did you have a busy week?	Yes, I did.	No, I didn't.
1. Did you have to work late?	✓	☐
2. Did you write a report or paper?	☐	☐
3. Did you have a lot of appointments?	☐	☐
4. Did you make a lot of phone calls?	☐	☐
5. Did you go out a lot in the evening?	☐	☐

> *About you*

B ***Pair work*** Ask and answer the questions. How many things do you have in common?

"Did you have to work late?"   *"Yes, I did. I had to work late on Monday and Tuesday."*

3 *Grammar* *Simple past yes-no questions and short answers*

Did you **go out** a lot last week?
　Yes, I **did**. I went to a movie and a party.
　No, I **didn't**. I didn't go out a lot.

Did you **play** tennis last weekend?
　Yes, I **did**. I played tennis last Sunday.
　No, I **didn't**. I didn't play tennis.

> **Past time expressions**
>
> | | |
> |---|---|
> | last night | last week |
> | yesterday | last month |
> | two days ago | last year |
> | last Friday | |

A Unscramble the words to make questions. Then ask and answer the questions with a partner. Try to remember your partner's answers.

1. last night / go to bed / did / you / late ?　　*Did you go to bed late last night?*

2. read / did / a lot of / you / last summer / books ?　　_____

3. your family / a trip / did / last year / take ?　　_____

4. get together / did / you and your friends / last Friday ?　　_____

5. a concert / go to / did / last month / you ?　　_____

6. you / meet / last year / any new people / did ?　　_____

> *About you*

B ***Pair work*** Find another partner. Ask and answer questions about your first partners. How much do you remember?

"Did Ichiro go to bed late last night?"   *"Yes, he did. He went to bed at 2:00 a.m."*

4 *Vocabulary notebook* *Ways with verbs*

See page 106 for a new way to log and learn vocabulary.

1 Conversation strategy *Appropriate responses*

A Complete these conversations. Choose a response from the box.

A **I have an exam tomorrow.**
B _____

A **It's my birthday today.**
B _____

> **Congratulations!**   **Thank goodness!**
> **Happy birthday!**   **Good luck!**

Now listen. Why is Eve so exhausted?

Eve Thank goodness it's Friday. I'm exhausted! I had exams all week.

Mark You did? You poor thing!

Eve Then today I took my driver's test.

Mark Finally! How did you do?

Eve I passed.

Mark You did? Congratulations!

Eve Thanks. It's a nice birthday present.

Mark It's your birthday? Happy birthday! Do you have any plans?

Eve Well, I have an interview tonight at the hospital – I want to volunteer there.

Mark Good for you. Well, good luck with the interview.

Notice how Mark responds to Eve's news. He uses expressions like *You poor thing*. Find examples in the conversation.

"You poor thing!"

B Think of responses to these comments. Use the ideas on the right. Practice with a partner.

1. I'm twenty-one today!
2. I had an interview last week, and I got the job!
3. My neighbor's sick, so I did her shopping today.
4. My football team has a big game on Friday.
5. Today's our wedding anniversary.
6. I have a really bad cold.

Good luck!

Happy birthday!

Congratulations!

Good for you!

You poor thing!

"I'm twenty-one today!"   *"Happy birthday!"*

2 *Strategy plus* You did?

You can say
You did? to show
that you're interested
or surprised, or just
that you're listening.

"I passed my driver's test."

"You did?"

In conversation . . .

You can also say **Did you?**
to show that you're listening.

A Complete the conversations with *You did?* and then add a question.
Practice with a partner. Continue each conversation.

1 A I had a nice, relaxing day at the beach last weekend.
   B <u>You did</u> ? <u>Did you go swimming</u> ?

2 A I had four exams this week – I had three yesterday.
   B _____ ? _____ ?

3 A I had a busy day today. I had 50 e-mails this morning.
   B _____ ? _____ ?

A *I had a nice, relaxing day at the beach last weekend.*

B *You did? Did you go swimming?*

A *No, but I went windsurfing. . . .*

About you → **B** *Pair work* Talk about the things you did last week.
Respond with *You did?* How long can you continue
your conversation?

3 *Listening* What a week!

A 🔘 Listen. What kind of week did these people have? Check (✔) the correct word.

1 George
☐ terrible
☐ busy
☐ relaxing

2 Karen
☐ exciting
☐ nice
☐ terrible

3 Brittany
☐ boring
☐ fun
☐ awful

B 🔘 Listen again. Choose the best response to give each person.

1. George _____
2. Karen _____
3. Brittany _____

a. Oh, good. Thank goodness for that!
b. You did? Good for you!
c. You did? You poor thing!

4 *Free talk* Yesterday . . .

For more speaking practice, go to the back of the book.
Student A: See *Free talk 10A*. Student B: See *Free talk 10B*.

1 Reading

A Do you or your friends keep a journal? What topics do people write about in journals? Add other ideas.

problems   family   school

B Read Ashley's journal entries for last week. What topics did she write about?

Monday

I started my new job at the design company. I like it a lot. I met Brad after work. He wants to go out for dinner on Wednesday, but I already made dinner plans with some people from work, and I really want to go. Anyway, Brad never pays!

Tuesday

Brad called me three times at work. I turned off my cell phone in the end! I met Rachel for lunch. She works in another department. She's funny. We laughed a lot.

Thursday

I had a really busy day! I went to a big meeting, and it went on all day. I came home late. I called Brad before I went to bed, but he didn't answer. I'm sure he's mad at me.

Friday

I had a great day. We finished work early today. (My new boss is nice!) Then I met Brad for dinner in the evening. We argued about Wednesday – the night I went out with friends from work. I don't think I want a boyfriend right now.

Saturday

Actually, I'm writing this on Sunday morning! I felt too tired when I came home last night. I went to a party, and I met some really fun people. A guy called José wanted my phone number. I hope he calls!

C Read the journal entries again. Do you agree with these statements about Ashley? Why or why not? Compare with a partner.

	Agree	Disagree
1. Ashley enjoyed her first day at her new job.	☐	☐
2. She wanted to go out with Brad on Wednesday.	☐	☐
3. She's not happy with Brad.	☐	☐
4. She didn't have fun at the party.	☐	☐
5. She doesn't want a boyfriend right now.	☐	☐
6. She had a busy week.	☐	☐

2 Writing *A journal*

A Write a journal for two days last week. What did you do? Write a short paragraph about each day.

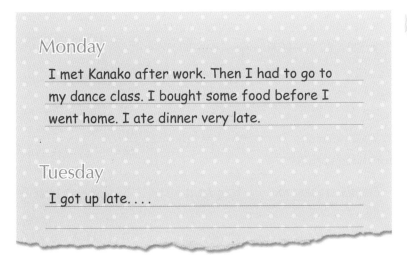

Monday

I met Kanako after work. Then I had to go to my dance class. I bought some food before I went home. I ate dinner very late.

Tuesday

I got up late. . . .

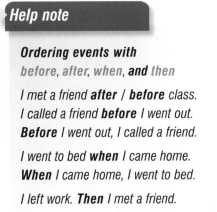

Help note

Ordering events with
before, after, when, **and** *then*

I met a friend **after** / **before** *class.*
I called a friend **before** *I went out.*
Before *I went out, I called a friend.*

I went to bed **when** *I came home.*
When *I came home, I went to bed.*

I left work. **Then** *I met a friend.*

B *Pair work* Read your partner's journal. Then find out more about your partner's week. Ask questions.

"So, you met Kanako after work on Monday. Did you go out for coffee?"

3 Listening and speaking *Don't forget!*

A *Pair work* How do you remember the things you have to do?
Ask and answer the questions.

❶ *Do you write lists?* ❷ *Do you have a daily or weekly planner?* ❸ *Do you write notes?* ❹ *Do you just try to remember everything?*

B Listen. How do these people remember things? Match the people and what they do.

1. Angela _____   a. puts notes on the refrigerator.
2. Kevin _____   b. tries to remember things without a list.
3. Sun Yee _____   c. writes a list every week.
4. Rafael _____   d. has a daily planner.

About you **C** *Class activity* How do your classmates remember things? Ask them the questions in A above. How many students say yes to each question?

Vocabulary notebook

Ways with verbs

Learning tip *Making notes on verbs*

When you write down a new verb, make notes about it. Is it regular or irregular?
How do you spell the different forms? How do you pronounce the endings?

watch (R)	watches /IZ/	watching	watched /t/
take (IR)	takes /s/	taking	took

1 Can you complete the chart for these verbs?

	Regular or irregular?	Simple present for *he*, *she*, *and it*	*-ing* form	Simple past
1. study	regular	studies /z/	studying	studied /d/
2. chat				
3. invite				
4. do				
5. buy				
6. meet				

2 Here are the simple past forms of some irregular verbs. Can you figure out the base forms? Complete the chart.

eat	ate		felt		made		sang		thought
	bought		forgot		meant		sat		told
	brought		found		met		saw		took
	came		gave		paid		sent		went
	chose		got		put		slept		went out
	cost		got up		put on		sold		woke up
	did		had		ran		spent		won
	drank		knew		read		spoke		wore
	drove		left		said		swam		wrote

On your own

Before you go to sleep tonight,
think of all the things you did today.
How many things can you remember?

I ate a big dinner.
I watched TV.
I...zzzZZZz

Looking back

In Unit 11, you learn how to . . .

■ use the past of *be*.

■ ask simple past information questions.

■ talk about past experiences.

■ use expressions with *go* and *get*.

■ show interest in other people.

■ use *Anyway* to change the topic or end a conversation.

Before you begin . . .

What do you remember about these things?

■ my first pet ■ my first friend ■ my first home

What other "firsts" do you remember?

My first . . .

THE DAILY HERALD

I remember my first day of . . .

school

Jeff Chang's kindergarten class

Jeff Chang
" It was awful! I was so scared of the teacher. I remember her name was Ms. Johnson and that she was very strict. The other kids weren't too happy, either. We were all very quiet that day! "

work

Restaurant where Rosa Leon had her first job

Rosa Leon
" I had a part-time job in a restaurant. I was a server. I was young – only 16. Things were really busy that first day, so I was nervous! I made a lot of embarrassing mistakes, and my boss wasn't too pleased. But most people were nice because I was new. "

1 Getting started

A Listen. Why was Jeff scared? Why was Rosa nervous?

| Figure it out |

B Can you complete the answers to these questions about Jeff and Rosa? Then ask and answer the questions with a partner.

1. *A* Was Jeff's teacher strict?
   B Yes, she _____ very strict.

2. *A* Were Jeff and his classmates noisy?
   B No, they _____ noisy.

3. *A* Was Rosa's boss happy about her mistakes?
   B No, he _____ too pleased.

4. *A* Were Rosa's customers nice?
   B Yes, they _____ nice because Rosa was new.

2 Grammar *Simple past of be* 💿

I	**was** only 16.	I	**wasn't** very old.	**Were** you nervous?	
You	**were** nervous.	You	**weren't** relaxed.		Yes, I **was**. / No, I **wasn't**.
She	**was** strict.	She	**wasn't** very nice.	**Was** it fun?	
It	**was** awful.	It	**wasn't** fun.		Yes, it **was**. / No, it **wasn't**.
We	**were** quiet.	We	**weren't** noisy.	**Were** they nice?	
They	**were** scared.	They	**weren't** happy.		Yes, they **were**. / No, they **weren't**.

wasn't = was not   *weren't = were not*

A Complete these conversations with *was, wasn't, were,* or *weren't*. Practice with a partner.

> ⸬⸬⸬ **In conversation . . .**
>
> **Was** is one of the top 20 words.

1. *A* Do you remember your first teacher?
   B Yeah, Mr. Davis. He _____ a lot of fun.
      He _____ never strict. _____ your teachers fun?
   A No, they _____ . They _____ always very strict.

2. *A* Do you remember the first CD you bought?
   B Yeah, it _____ The Backstreet Boys.
      They _____ my favorite group.

3. *A* Tell me about your first best friend.
      _____ you in school together?
   B No, we _____ . She _____ in my class.
      She _____ my neighbor. Our parents _____ friends.

4. *A* Did you have a pet when you _____ a kid?
   B Yeah. My first pet _____ a little pony. It _____ very big.

About you → **B** *Pair work* Ask and answer the questions. Give your own answers.

3 Speaking naturally *Stress and intonation*

Were you **nér**vous?   No, I/**was**n't.   I was re**lax**ed.

A 💿 Listen and repeat the sentences above. Notice how the voice falls or rises on the stressed words.

B 💿 Now listen and repeat these questions and answers.

Do you remember your first English class?

1. *A* Was the class **eas**y?
2. *A* Were the other students **good**?
3. *A* Were they **nice** to you?
4. *A* Was your teacher **strict**?

   B No, it **was**n't. It was **hard**!
   B Yes, they were all very **smart**.
   B Yes, they **were**. They were very **friend**ly.
   B Yes, she **was**. But she was **nice**.

About you → **C** *Class activity* Interview three students about their first English class. Ask the questions above.

109

1 Building language

A Where are good places to go on vacation? Tell the class.

B  Listen. What did Jason do on his vacation?
Practice the conversation.

Diana Great picture! When did you get back?
Jason Last night.
Diana So how was your vacation?
Jason Oh, it was wonderful.
Diana Where did you go exactly?
Jason We went to Hawaii.
Diana Wow! What was the weather like?
Jason It was hot, but not too hot.
Diana Nice. So, what did you do there?
Jason We went to the beach every day, and I went
         parasailing. I didn't want to come home.
Diana Well, I'm glad you did. . . . I have a ton
         of work for you!

Figure it out

C Can you complete these questions? Then ask a partner.

1. What _____ you _____ last summer?
2. What _____ the weather like?

2 Grammar *Simple past information questions*

How was your vacation?	It was fun.	**Where** did you go?	To Hawaii.
Where were you exactly?	In Hawaii.	**Who** did you go with?	A couple of friends.
How long were you there?	A week.	**What** did you do?	We went to the beach.
What was the weather like?	It was hot.	**When** did you get back?	Last night.

About you

Write questions about a vacation for these answers. Then ask and answer the
questions with a partner. Give your own answers.

1. <u>How was your last vacation</u>_____ ? It was great.
2. _____ ? I went to England.
3. _____ ? Awful. It rained every day.
4. _____ ? My best friend.
5. _____ ? Two weeks.
6. _____ ? We saw Buckingham Palace.

"How was your last vacation?"   *"It was OK. It was very short."*

3 *Building vocabulary*

A 💿 Listen to these memories of trips. Match the memories with the pictures.

❶ "I **went hiking** with a friend in Peru, and we **got lost**."

❷ "I **got** a new camera from my mom for my trip to Africa."

❸ "I **got sick** on our honeymoon, right after we **got married**."

❹ "I **went on a trip** across Canada with a friend. It was awful. We didn't **get along**."

❺ "I **went to see** a band in Miami. I met the lead singer, and I **got his autograph**."

❻ "I **went snorkeling** in Thailand. It was great, but I **got a bad sunburn**."

Word sort

B Make word webs for *get* and *go* with expressions from the sentences above. Add ideas.

go (went) — go hiking

get (got) — get lost

About you

C *Pair work* Talk about a time you did one of these things. Ask questions to find out more.

- got lost
- got scared
- got home very late
- went on a trip
- went camping

"Last fall I went on a road trip with a friend."   *"You did? Where did you go? . . . Who drove?"*

4 *Vocabulary notebook* *Past experiences*

See page 116 for a new way to log and learn vocabulary.

Lesson C How was your weekend?

1 Conversation strategy *Answer a question; then ask a similar one.*

A **Can you add a question to continue this conversation?**

A ***How was your weekend?***
B ***Good, thanks. I went to a wedding.*** ______________________ ?

Now listen. How was Jessica's weekend? How was Ben's weekend?

Ben ***So, how was your weekend, Jessica?***
Jessica ***Great! Gina and I went biking out in the country.***
Ben ***Oh, really?***
Jessica ***Yeah, it was fun, but there were lots of hills. I was exhausted by the end of the day.***
Ben ***Yeah, I bet.***
Jessica ***So. . . . Anyway, what did you do?***
Ben ***Oh, I had a party Saturday. It was good.***
Jessica ***Really? Nice.***
Ben ***Well, anyway, . . . I have to go. I have a meeting now. See you later.***

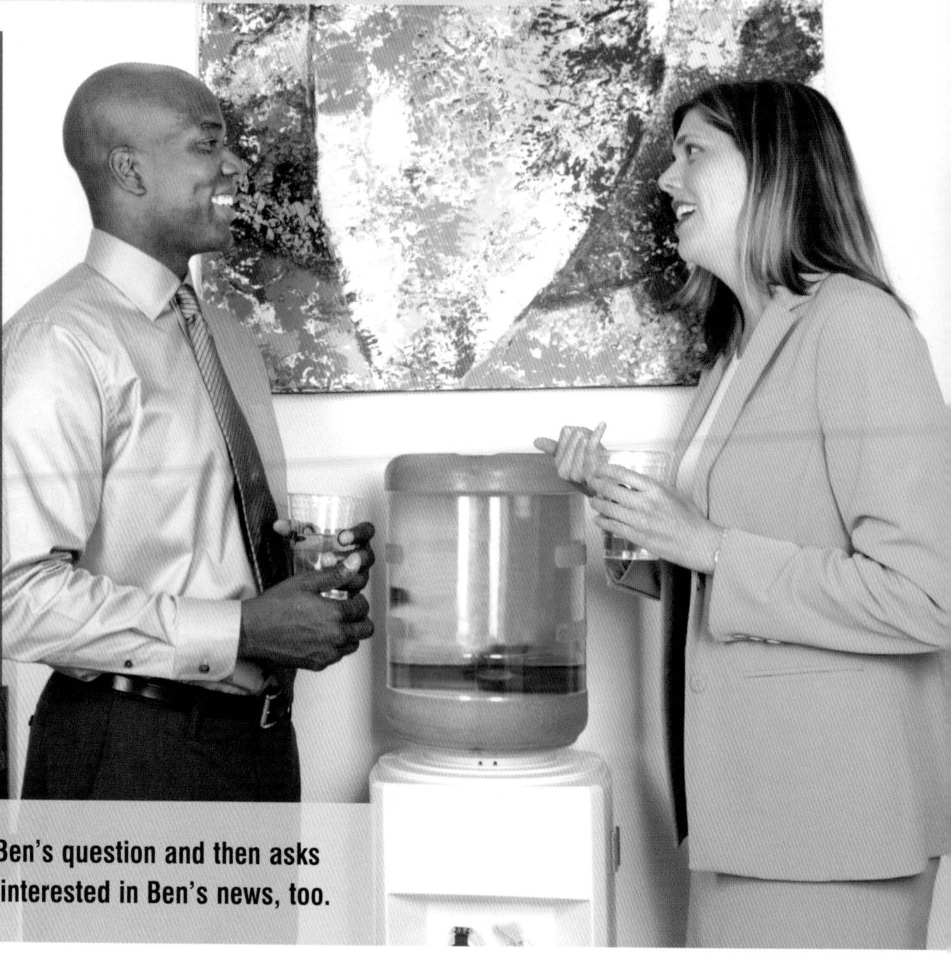

Notice **how Jessica answers Ben's question and then asks a similar one. She shows she is interested in Ben's news, too.**

About you → **B** Answer each question. Then think of a similar question to ask. Practice your conversations with a partner.

1. *A* How was your weekend? Did you have a good one?
   B *Answer:* ______________________
   Then ask: ______________________

2. *A* Did you do anything fun on Friday night?
   B *Answer:* ______________________
   Then ask: ______________________

3. *A* What did you do on Sunday?
   B *Answer:* ______________________
   Then ask: ______________________

SELF-STUDY AUDIO CD
CD-ROM

2 Strategy plus *Anyway*

You can use **Anyway** to change the topic of a conversation.

"Anyway, what did you do?"

You can also use **Anyway** to end a conversation.

"The party was good. Well, anyway, . . . I have to go."

In conversation . . .

Anyway is one of the top 300 words.

Why are these people saying *anyway*? Circle **a** or **b**. Then practice with a partner.

1. *A* How was Saleem's party last weekend?
   B Good. He cooked some great food. **Anyway**, do you still want to go out tonight?

   a. to change the topic
   b. to end the conversation

2. *A* Let's go camping together one weekend.
   B That sounds nice. **Anyway**, call me later, and we can talk about it.

   a. to change the topic
   b. to end the conversation

3. *A* Yes, we had a lot of fun on Saturday. **Anyway**, I forgot to tell you about my new car.
   B Oh, what's it like?

   a. to change the topic
   b. to end the conversation

4. *A* I really enjoyed that movie.
   B Yeah, me too. Well, **anyway**, it's getting late. I have to go. See you tomorrow.

   a. to change the topic
   b. to end the conversation

3 Listening and speaking *Weekend fun*

A Listen to two friends talk about their weekend. Circle the topics.

baseball   **biking**   **a party**   **the beach**   **hiking**

B Listen again and answer the questions.

1. What time did the man get to Simon's place on Saturday?
2. When did the woman leave Simon's place?
3. Where was the woman on Sunday? Did she have fun?
4. What did the man do on Sunday? What was the weather like?

 About you

C *Class activity* Have a conversation with a classmate about last weekend. Then end your conversation, and talk with another classmate. Talk to at least three people.

4 Free talk *Guess where I went on vacation.*

See *Free talk 11* for more speaking practice.

A funny thing happened . . .

1 Reading

A Do you ever read the letters people send in to magazines? What topics do people write about? Add ideas. Tell the class.

problems   **health**   **personal stories**

B Read the letter to a magazine. What is it about? What happened to Alexa?

Letters from our readers

Last week we asked you to send in stories about an unforgettable experience. Reader Alexa Astor wins a weekend for two at the Sun Valley Spa for this letter.

Dear *City Life*:

A funny thing happened last Saturday afternoon. I went to the mall to meet my friend Sammy. I was a bit early, so I decided to have a snack and a drink.

I went to a new café called The Metro. It was a little expensive, so I just got a soda. It was really crowded, but I found a table and sat down.

Then a guy came over and said, "Is this seat free?" He was gorgeous, so I said, "Sure." Anyway, he had a cup of coffee and a sandwich. He drank the coffee and ate half of the sandwich, and then he left. I was hungry, so I ate the other half.

Then a few minutes later, he came back! He was on his cell phone, so I didn't explain about the sandwich. I just left. I was so embarrassed!

But things got worse. I met my friend Sammy about 15 minutes later, and she said, "Let's go and meet my cousin Josh. He just called me from The Metro Café."

And, yes, it was the same guy!

Continued on next page.

C Read the letter again. Then match the two parts of each sentence.

1. Alexa got to the mall early before her friend Sammy, so __c__
2. The café was expensive, so ____
3. The café was very crowded, so ____
4. The guy ate half of his sandwich, and then ____
5. Alexa ate the rest of the guy's sandwich because ____
6. When the guy came back to the table, ____
7. Later Alexa met Sammy's cousin, and ____

a. he left the café.
b. she only bought a drink.
c. she decided to go to a café.
d. she got embarrassed and left.
e. she was hungry.
f. it was the guy from the café.
g. a guy sat down at her table.

2 *Writing* He said, she said

A Look at the rest of Alexa's letter. Alexa writes about what people said – she includes "quotations." Correct the punctuation. Use the information in the note.

> ### Letters from our readers
> *Continued from previous page.*
>
> S   "A   J
> $\cancel{s}$ammy said, $\cancel{a}$lexa, this is my cousin $\cancel{j}$osh."
>
> he said how are you
>
> I said hello
>
> he said we met a few minutes ago
>
> sammy said oh good you know each other
>
> so are you hungry
>
> I said no I'm not
>
> he said yes I am I'm very hungry

> ### Help note
>
> **Punctuation with speech**
> - Use quotation marks (" ") around the things people say.
> - Use a comma (,) after **said**.
> - Use a capital letter to start a quotation.
>   > A guy said, **"Is this seat free?"**
>   > I said**, "Sure."**

B What did they say next? Write six sentences to finish the story. Then read your ending to the class.

3 *Listening and speaking* Funny stories

A Listen to Miranda and John tell part of a story. Circle the correct information.

Miranda
I did something really embarrassing about a month ago. . . .

John
I said something once to a dinner guest. . . .

1. Miranda was **at work / in a store**.
2. Her friend **loves / hates** shopping.
3. They looked at a **dress / sweater**.
4. Miranda **liked / didn't like** the colors.

1. John was **10 / 20** years old.
2. His father's **boss / friend** came for dinner.
3. John and the man talked about **school / work**.
4. John **liked / didn't like** his new teacher.

B Now choose the best ending for each story. Then listen to the whole story, and check your guesses.

1. Miranda's story
   a. Then my friend said, "Actually, I bought one last week."
   b. The clerk said, "Do you like this season's colors?"

2. John's story
   a. My teacher said, "You look tired. Were you up late last night?"
   b. My teacher said, "I hear you met my father last night."

C *Pair work* Retell one of the stories above to a partner, or tell a funny story of your own.

Learning tip *Time charts*

You can use a time chart to log new vocabulary.

1 Complete the sentences on the time chart below with the correct verbs.
You can use a verb more than once.

bought	had	took	didn't have	went
got	✓ lived	was	didn't get along	

15 years ago	My family __lived__ in Hawaii.
10 years ago	I _____ in high school.
5 years ago	I _____ my driver's license and _____ my first car.
2–4 years ago	I _____ my first trip abroad.
last year	I _____ sick and _____ in the hospital for two weeks.
last month	My brother _____ married and _____ to Fiji on his honeymoon.
last week	My friend Jo _____ a party. It _____ boring. I _____ a good time.
last weekend	I _____ hiking with a friend. It was awful – we _____ .

2 Make a time chart about your past experiences.

_____ years ago	_____
_____ years ago	_____
_____ years ago	_____
_____ years ago	_____
last year	_____
last month	_____
last week	_____
yesterday	_____

On your own

Make a time chart, and put it on your wall.
Look at it every day.

Last week: I started a new job.
Last month: I was on vacation.

Fabulous food

In Unit 12, you learn how to . . .

■ use *many* and *much* with countable and uncountable nouns.

■ use *some* and *any* in statements and questions.

■ use *would like* for offers and requests.

■ talk about favorite foods and eating habits.

■ use *or something* and *or anything*.

■ add *or . . . ?* to *yes-no* questions.

Before you begin . . .

Can you find these foods in the pictures? Which foods did you eat yesterday?

■ milk, cheese, and eggs ■ fruit and vegetables ■ meat: beef and chicken

■ seafood: fish and shellfish ■ bread, rice, and pasta

Eating habits

Kayla Hi, Mom and Dad!
I need some help fast!
I invited some friends
for dinner tonight, and I
don't know what to cook.

Andrea's a vegetarian,
so she doesn't eat
meat, fish, cheese,
or eggs. I guess she
just eats a lot of fruit
and vegetables,
and maybe rice.

Colin's on a diet.
He can't eat much
rice, bread, or pasta.
But he eats a lot of
meat, cheese, eggs,
and vegetables, like
carrots and cucumbers.

And James is picky –
I mean, he doesn't
eat many vegetables.
And he's allergic
to milk and shellfish.
But he likes potatoes.
Oh, and bananas.

Please call me! Bye.

1 Getting started

A Listen. Kayla is leaving a phone message for her parents. What is her problem?
Which plate of food is right for Andrea? for Colin? for James?

Figure it out

B Find the food words in Kayla's message. Which are singular? Which are plural?
Write them in the chart.

Singular			Plural	
meat			eggs	

About you

C *Pair work* Which of the foods above do you like? Which don't you like?
Tell your partner.

"I like meat. How about you?"   *"Um, I don't eat meat, but I really like fish."*

2 Grammar *Countable and uncountable nouns*

Countable nouns:	**Uncountable nouns:**
Use a/an or plural -s.	**Don't use a/an or plural -s.**
I have **an egg** for breakfast every day.	I drink **milk** every morning.
I don't eat **bananas**.	I don't eat **seafood**.
How many eggs do you eat a week?	**How much** milk do you drink a day?
I eat **a lot of** eggs.	I drink **a lot of** milk.
I don't eat **many** (eggs).	I don't drink **much** (milk).
I don't eat **a lot of** eggs.	I don't drink **a lot of** milk.
Examples: vegetables, potatoes	*Examples:* cheese, meat, fish

A Circle the correct words in these questions and answers. Then practice with a partner.

1. *A* How **much / many** fruit do you eat a week?
    B Well, I have **orange / an orange** every day for breakfast,
        and I eat **a lot of** / **much** fruit after dinner for dessert.

2. *A* How often do you eat **vegetable / vegetables**?
    B I usually eat **many / a lot of** French fries. Is that a vegetable?

3. *A* How **much / many** times a week do you eat **rice / rices**?
    B About twice a week. But I eat **potato / potatoes** every day.

4. *A* Do you eat **many / a lot of** seafood?
    B Well, I eat **much / a lot of** fish, but I can't eat **shellfish / a shellfish**.

5. *A* Do you eat **meat / meats**?
    B Well, I don't eat **beef / beefs**, but I eat **many / a lot of** chicken.

6. *A* How **much / many** eggs do you eat a week?
    B I don't eat **much / many**. I don't really like **egg / eggs**.

About you → **B** *Pair work* Ask and answer the questions. Give your own answers.

3 Talk about it *What's your daily diet?*

Group work Discuss the questions. Do you have similar habits?
Tell the class one interesting thing about a person in your group.

  ▶ Are you a picky eater? What foods do you hate?
  ▶ Are you allergic to any kinds of food? What kinds?
  ▶ Are you on a special diet? What can't you eat?
  ▶ How many times a day do you eat?
     Do you ever skip meals?
  ▶ In your opinion, what foods are good for you?
     What foods aren't?
  ▶ Do you have any bad eating habits? What are they?

What's for dinner?

1 Building vocabulary

A Listen and say the words. Which foods do you like? Which don't you like? Tell the class.

spinach   onions   peppers   lettuce   tomatoes   oil   melon   apples
          garlic   green beans   butter   mangoes   strawberries   pineapple

coffee   sugar   tea   cereal   potato chips   peanuts   ice cream   cookies   lamb   salmon   shrimp   hamburger meat

Word sort

B What foods do you regularly buy? Complete the chart. Compare with a partner.

We buy a lot of . . .	We don't buy much . . .	We don't buy many . . .	We never buy . . .
melon			

2 Building language

Listen. What does Dan want for dinner?
Practice the conversation.

Kathy What do you want for dinner tonight?
Dan I don't know. Would you like to go out?
Kathy No, we eat out all the time. I'd like to stay home tonight.
Dan OK. Um . . . I think I'd like some chicken. Do we have any in the freezer?
Kathy Uh . . . no, we need to get some. And we don't have any vegetables, either.
Dan So, I guess we have to go to the grocery store.
Kathy Hmm. I have another idea. Let's just go out for dinner!

3 Grammar *Would like; some and any*

Would you like to go out?
   No, I'd like to stay home.

What would you like?
   I'd like some chicken.

Would you like some tea?
   Yes, please. / No, thanks.

Do we have any chicken?
   Yes, we have some (chicken).
   No, we don't have any (chicken).

Do we have any vegetables?
   Yes, we have some (vegetables).
   No, we don't have any (vegetables).

In conversation . . .

Any is common in questions:
   *Do you have **any** cookies?*
Some is common in questions
that are offers or requests:
   *Would you like **some** chicken?*
   *Can I have **some** chocolate?*

A Complete the questions and answers with *some* or *any*.
Then practice with a partner.

1. *A* I'm sleepy. Would you like to get _____ coffee after class?
   B I just had _____ before class, but I can go with you and get something else.

2. *A* I'm hungry. Do you have _____ chocolate or candy with you?
   B No, but I have _____ peanuts. Would you like _____ ?

3. *A* How many snacks do you eat a day?
   B Actually, I don't eat _____ . I don't eat between meals.

4. *A* I have _____ cookies in my backpack. Would you like _____ ?
   B No, thanks. I don't want _____ right now. But can I have _____ later?

About you → **B** *Pair work* Ask and answer the questions. Give your own answers.

A **I'm sleepy. Would you like to get some coffee after class?**
B **Sure. Where would you like to go?**

4 Speaking naturally *Would you . . . ?*

What would you *like?*   *Would you* like a snack?   *Would you* like to have dinner?

A Listen and repeat the questions above. Notice
the pronunciation of *Would you . . . ?*

B Listen and complete the questions. Then use
the questions to make dinner plans with a partner.

1. What would you like to _____ ?
2. Would you like to _____ ?
3. Would you like to _____ ?
4. Where would you like to _____ ?
5. What would you like to _____ ?

5 Vocabulary notebook *I love to eat!*

See page 126 for a new way to log and learn vocabulary.

Let's take a break for lunch.

1 Conversation strategy *or something and or anything*

A Can you use *or something* and *or anything* to complete these sentences?

 A **What do you want for lunch? I'd just like a snack _____ .**

 B **Me too. I don't want a big meal _____ .**

Now listen. What do Emily and Matt decide to do for lunch?

Emily **Let's take a break for lunch.**

Matt **Sure. Would you like to go out or . . . ?**

Emily **Well, I just want a sandwich or something.**

Matt **OK. I don't want a big meal or anything, either. But I'd like something hot.**

Emily **Well, there's a new Spanish place near here, and they have good soup.**

Matt **That sounds good.**

Emily **OK. And I can have a sandwich or a salad or something like that.**

Matt **Great. So let's go there.**

Notice how Emily and Matt use *or something* (like that) and *or anything*. They don't need to give a long list of things. Find examples in the conversation.

"I don't want a big meal or anything."

B Complete the questions and answers with *or something* and *or anything*. Then practice with a partner.

① *A* Do you eat lunch every day?

 B Yeah, I usually have a salad _____ and some fruit.

② *A* What do you have for breakfast usually?

 B Oh, I have some yogurt and a banana _____ .

 A You don't have eggs _____ ?

③ *A* Do you have any water _____ ? I'm thirsty.

 B No, but would you like to go out for a soda _____ ?

 A Yeah, we can get a muffin or a cookie _____ , too.

About you → **C** *Pair work* Ask and answer the questions. Give your own answers.

SELF-STUDY
AUDIO CD
CD-ROM

2 *Strategy plus* or . . . ?

You can use *Or . . . ?*
at the end of *yes-no*
questions to make
them less direct.

**Would you like to
go out or . . . ?**

**Well, I just want a sandwich
or something.**

About you → **Pair work** ·Check (✔) the questions you can end with *or . . . .*
Then ask and answer all of the questions.

In conversation . . .

Or is one of the top 50 words.

- ✔ 1. Do you go out for lunch every day _or . . ._ ?
- ☐ 2. Which restaurants around here are good for lunch _____ ?
- ☐ 3. Do you like to have something light _____ ?
- ☐ 4. What did you have for lunch yesterday _____ ?
- ☐ 5. Do you like to have lunch alone _____ ?
- ☐ 6. Who do you usually have lunch with _____ ?
- ☐ 7. Do you ever make your own lunch _____ ?
- ☐ 8. Do you usually have lunch around 1:00 _____ ?

A *Do you go out for lunch every day or . . . ?*
B *Well, I usually bring my lunch, but today I didn't.*

3 *Listening and speaking* Lunchtime

A 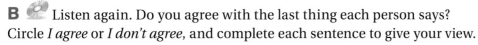 Listen to the conversations, and match the two parts of each sentence.

1. Rex _____   a. doesn't want anything to eat.
2. Amy _____   b. wants a big meal.
3. Omar _____   c. would like something hot.
4. Gemma _____   d. just wants a drink.

B Listen again. Do you agree with the last thing each person says?
Circle *I agree* or *I don't agree*, and complete each sentence to give your view.

1. **I agree / I don't agree**. I like to _____ .
2. **I agree / I don't agree**. I think _____ .
3. **I agree / I don't agree**. I usually _____ .
4. **I agree / I don't agree**. I guess _____ .

C *Pair work* Make up your own conversation about lunch plans.
Act out your conversation for the class.

4 *Free talk* Do you live to eat or eat to live?

See *Free talk 12* for more speaking practice.

Great places to eat

1 Reading

A What makes a good restaurant? Check (✓) the three things that are most important to you when you go out to eat. Tell the class.

A good restaurant has . . .
- ☐ a nice atmosphere.
- ☐ live music.
- ☐ excellent food.
- ☐ low prices.
- ☐ good service.
- ☐ wonderful desserts.

B Read the restaurant guide. Choose a restaurant you would like to try. Tell a partner why you'd like to go there.

★ ★ ★ ★ **RESTAURANT GUIDE**   International Restaurants   25

EL PATIO

Enjoy a Latin American night out – a fiesta of fun, music, and authentic Mexican food. We recommend the great seafood and chicken dishes. Ask to sit on the patio under the stars, and listen to a mariachi band while you eat.

Food: ★ ★ ★   *Service:* ★ ★   *Price:* $$

MAMMA MIA

If you'd like a cheap night out, then this is a great Italian place for pasta, pizza, and salad. Try their delicious home-made ice cream. But don't come here looking for a quiet place to talk – it's a very popular place for students to hang out on the weekends.

Food: ★ ★   *Service:* ★   *Price:* $

MEKONG

Would you like to try something different? Try the menu at this busy little Vietnamese restaurant. We recommend the sticky rice and beef.

Food: ★ ★ ★   *Service:* ★ ★   *Price:* $$

PARIS

If you're planning a quiet dinner for two in a romantic atmosphere, try Paris. This restaurant has fantastic French cuisine – expensive but great for special occasions.

Food: ★ ★ ★ ★   *Service:* ★ ★ ★ ★   *Price:* $$$$

SAKURA

This quiet and friendly restaurant serves the best sushi in town. The service is excellent, and you can watch the chef prepare your meal.

Food: ★ ★ ★   *Service:* ★ ★ ★ ★   *Price:* $$$

STIR CRAZY

Here's something new! At Stir Crazy, you make your own dinner. Fill a bowl with vegetables, tofu, rice, or noodles, and add some shrimp, beef, or chicken. Then cook it at your table. All you can eat for $10.

Food: ★ ★   *Service:*   *Price:* $

SYLVESTER'S STEAK HOUSE

It's noisy, expensive, and crowded, but Sylvester's is *the* place to go for steak. We recommend it!

Food: ★ ★ ★   *Service:* ★   *Price:* $$$

C Read the article again, and answer these questions. Compare your answers with a partner.

Which restaurant do you think . . .

- ▦ has the best atmosphere?
- ▦ sounds like fun?
- ▦ sounds like a good place for a special dinner?
- ▦ sounds like a place to "hang out" with your friends?
- ▦ you would like to go to with your family?
- ▦ is not worth trying?

2 *Listening and writing* *Do you recommend it?*

A 🔘 Listen to Dave talk about a restaurant he went to last week. What do you find out about it? Circle the correct words.

1. The restaurant was **Italian / Indian**.
2. They have great **seafood / chicken**.
3. It's **good / not good** for vegetarians.
4. He had **a steak / some fish**.
5. The service was **friendly / slow**.
6. The atmosphere was **formal / fun**.
7. He **recommends it / doesn't recommend it**.

B Write a review of a restaurant or café you know. Use the ideas above to help you. You can start and end like this:

THE GARLIC POT

Last week I went to a great restaurant. It was called The Garlic Pot. They serve excellent seafood and steaks, and every dish has garlic in it. . . .

. . . I highly recommend it.

> **Help note**
>
> **Useful expressions**
>
> | Was it . . . | good? | bad? |
> |---|---|---|
> | The restaurant was | good. | terrible. |
> | The service was | excellent. | slow. |
> | The servers were | friendly. | unfriendly. |
> | The meal was | delicious. | awful. |
> | The food was | tasty. | tasteless. |
> | The potatoes were | hot. | cold. |

C Read your classmates' reviews. Which restaurant would you like to try?

3 *Talk about it* *What are your favorite places to eat?*

Group work Discuss the questions. Do you have similar tastes?

- ► How often do you eat in restaurants?
- ► What kinds of restaurants do you go to?
- ► Do you have a favorite restaurant? Where is it? Why do you like it?
- ► What's the best restaurant in your neighborhood?
- ► Where can you get good, cheap food?
- ► Which restaurant don't you recommend? Why not?

I love to eat!

Learning tip *Grouping vocabulary*

You can group some vocabulary by the things you like and don't like.

1 Which of these types of food do you like? Which don't you like? Complete the word webs.

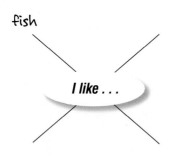

fish

I like . . .

✓cereal
✓fish
  fruit
  meat
  milk and cheese
  pasta and bread
  shellfish
  vegetables

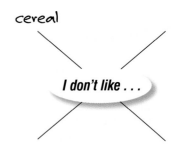

cereal

I don't like . . .

2 What foods do you love, and which do you hate? Complete the chart.

I love . . .	I like . . .	I don't like . . .	I can't stand . . .
			onions

On your own

Label your food at home in English. Learn the word before you eat the food!

Talk about food

The top food words people use with the verb *eat* are:

1.	meat	7.	vegetables
2.	beef	8.	seafood
3.	popcorn	9.	cheese
4.	eggs	10.	cookies
5.	fish	11.	pizza
6.	steak	12.	bread

apple

eggs

lettuce

1 What's the question?

Complete the conversation with information questions.
Then practice with a partner.

A So, <u>what did you do last night?</u>
B Last night? Oh, I went to see a band.
A You did? _____ ?
B The Travelers. They're a new band.
A Yeah? _____ ?
B They were great. We were there really late.
A _____ ?
B About 2:00 a.m. So anyway, _____ ?
A I just went home and made dinner. The usual.
B Well, let's go out tonight or something.
A Oh, OK. _____ ?
B Well, I'd like to see The Travelers again.
A OK. But let's leave before midnight.
B Sure. Then we can go see the midnight movie!

VAMPIRE STRIKES BACK
MIDNIGHT MOVIE

2 Do you have a balanced diet?

A How many words do you know for these categories of food? Complete the chart.

meat	seafood	vegetables	fruit	dairy	snacks
chicken				milk	

B Pair work Talk about the kinds of food you ate last week. Ask and answer questions
with *How much, How many, some,* or *any.*

"How much meat did you eat last week?"   *"Not much. I ate some chicken. Did you eat any meat?"*

3 Ask a question in two ways; answer and ask a similar question.

A Think of a *yes-no* question to add to each question below. End the question with *or . . . .*

1. What did you do last summer? I mean, <u>did you go away or . . .</u> ?
2. What would you like to do this summer? I mean, _____ ?
3. What did you do on your last birthday? I mean, _____ ?
4. How many times a week do you exercise? I mean, _____ ?

B Pair work Ask and answer the questions. After you answer a question, ask a similar one.

A *What did you do last summer? I mean, did you go away or . . . ?*
B *No, I didn't. I didn't take any vacation. Did you do anything special?*

4 *What's the right expression?*

Complete the conversation with these expressions. (Use *anyway* twice.)
Then practice with a partner.

or something	good luck	anyway	Good for you	You did
✓or anything	thank goodness	I know	Congratulations	You poor thing

Bryan How was your weekend? Did you go away __or anything__ ?

Julia No, but I went to a karaoke club.

Bryan _____ ? So, how was it?

Julia Great! I sang in a contest and won fifty dollars.

Bryan _____ ! I didn't know you were a singer.

Julia Well, I practiced every day for a month.

Bryan _____ !

Julia And _____ I practiced! Ten of my friends were there. So, _____ , did you do anything special?

Bryan Not really. I had to study for an exam on Saturday and Sunday.

Julia _____ ! You need to go out more.

Bryan Yeah. _____ . . . . Well, _____ , I have to go. I want to study my notes. But after the exam, let's meet for coffee _____ .

Julia OK. So, _____ with your exam.

5 *Show some interest!*

A Complete each sentence with a simple past verb. Then add time expressions to five sentences to make them true for you.

1. I _____ on a nice trip.
2. I _____ some new clothes.
3. I _____ someone famous.
4. I _____ an international phone call.
5. I _____ to a great party.
6. I _____ Italian food.
7. I _____ in the ocean.
8. I _____ English with a tourist.
9. I _____ some money.
10. I _____ lost in the city.

> I went on a nice trip last month.

B *Pair work* Take turns telling a partner your sentences. Respond with *You did?* and ask questions.

A **I went on a nice trip last month.**

B **You did? Where did you go? . . .**

Self-check

**How sure are you about these areas?
Circle the percentages.**

grammar
20% 40% 60% 80% 100%

vocabulary
20% 40% 60% 80% 100%

conversation strategies
20% 40% 60% 80% 100%

. . . .

Study plan

**What do you want to review?
Circle the lessons.**

grammar
10A 10B 11A 11B 12A 12B

vocabulary
10A 10B 11A 11B 12A 12B

conversation strategies
10C 11C 12C

1 Imagine you are a famous person. Make a name card.
Invent an e-mail address and phone number.

Name: Salma Hayek

E-mail address: salmah@cup.org

Telephone number: 885-555-7171

Name:

E-mail address:

Telephone number:

2 *Class activity* Now take turns introducing yourselves. Find three
famous people you like, and complete a name card for each one.

Name:

E-mail address:

Telephone number:

Name:

E-mail address:

Telephone number:

Name:

E-mail address:

Telephone number:

A Hello. I'm Salma Hayek.
B Oh, nice to meet you. I'm Ichiro Suzuki.
A I'm sorry. What's your name again?

1 Look at the things in this room. You have two minutes to memorize where things are.

2 *Pair work* Now close your books. What do you remember? Make a list of
things and where they are. Then open your books, and check your answers.

1. VCR – on the table
2. videos – on the VCR

A The VCR is on the table.
B Yes. And the videos are on the VCR.

1 Think of your favorite people. Write their names in the chart.

a TV star	an actor	a band	a family member

a friend	a singer	a writer	a sports star

2 Pair work Can you say three things about each person?
Score one point for each piece of information. Take turns.

A **Who's your favorite TV star?**
B **It's Jennifer Aniston. Her show is
  very funny. She's very smart.**
A **OK, that's two points for you.
  Now it's my turn.**

Scorecard		
	Me	My partner
a TV star		
an actor		
a band		
a family member		
a friend		
a singer		
a writer		
a sports star		
TOTAL		

Class survey Read the facts about the average New Yorker. Then ask three classmates if
they have the same habits. Write *Y* (yes) or *N* (no) in each box. Who is like a New Yorker?

What is the average New Yorker like?

Who is like a New Yorker?

	Student 1	Student 2	Student 3
70% of New Yorkers drink orange juice every day. _Do you drink orange juice every day?_			
53% read their horoscope every day. _____			
90% use an alarm clock. _____			
50% take their own food into the movie theater. _____			
80% sing in the car. _____			
78% eat lunch every day. _____			
66% eat cereal every week. _____			
71% listen to other people's conversations. _____			

A **Do you drink orange juice every day?**
B **No, I don't. I don't like orange juice.**
A **OK. Do you read your horoscope every day?**

1 Put a marker – an eraser, for example – on the start box.

2 Take turns tossing a coin. **Heads =** Move one space. **Tails =** Move two spaces.

3 Move around the board. Answer the questions and follow the instructions. Who gets to Hawaii first?

A *You go first, OK?*
B *Thanks. . . . Heads. I move one space.*
A *OK. Spell the teacher's full name.*
A *It's J-O-H-N E-V-A-N-S.*
B *No, it's J-O-N. Go back to class!*
A *Oh, no! . . . OK. Now it's your turn.*

Finish! HAWAII

14 What do people use computers for? Say up to six things.

6 things: Go to HAWAII and leave your laptop at home.
2 to 5: Stay here.
1: Go to a computer store at the MALL, and buy a laptop!

13 What *don't* you do on vacation? Say up to six things.

6 things: Go to HAWAII right now, and have a great vacation!
2 to 5: Take another turn.
1 or 2: Stay here.

 **12** the movie theater

9 Restaurant

10 Say the days of the week backwards. Start with *Saturday.*

Every day correct?
Yes: Take another turn.
No: Go to the MALL, and buy a new calendar!

11 What do you do in the evening? Say up to six things.

5 or 6: Go to the MOVIES.
3 or 4: Stay here.
1 or 2: Go to the RESTAURANT, and get a part-time job!

8 What's your favorite restaurant? Why? Say up to four things.

3 or 4: Go to the RESTAURANT for a free meal!
1 or 2: Stay here.

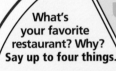 MALL

7 What's your best friend like? Say up to four things.

3 or 4: Stay here.
1 or 2: Go HOME and call your friend!

 **6**

5 How many kinds of TV shows do you know?

8+: Take another turn.
4 to 7: Stay here.
1 to 3: Go HOME and watch TV!

4 What do students buy for class? Say up to six things.

5 or 6: Go shopping at the MALL!
3 or 4: Stay here.
1 or 2: Go to CLASS and study!

 CLASS

Start here.

1 Spell the teacher's full name.

Correct: Take another turn.
Incorrect: Go back to CLASS.

2 How many expressions do you know for when you say good-bye?

5+: Say good-bye and go HOME.
1 to 4: Stay here.

 **3** HOME

Find the differences.

Pair work Look at the two neighborhoods below. How many differences do you see? Make a list.

A *There's a big park in Wilcox Hill.*
B *There's a park in Benson Valley, but it's small.*
A *OK. That's one difference.*

Differences

1. There's a big park in Wilcox Hill, but there's a small park in Benson Valley.

What's hot? What's not?

Group work Discuss these questions. Can you agree on three answers for each one?
Then compare with another group.

	Answer 1	Answer 2	Answer 3
What TV shows are popular right now? What is everybody watching?			
What bands are "hot"? Who are you and your friends listening to these days?			
What movies are playing this month? What movie stars are people talking about?			
What sports stars are in the news? Why? What are they doing?			
Where are all the cool people "hanging out" on weekends?			
What's happening in the news? What are people talking about?			

A **I think everybody is watching reality shows.**
B **Yes, reality shows are very popular.**
C **And people are also watching . . .**

How do you like to dress?

Class activity Find classmates who answer yes to the
questions. Write their names on the chart.

Find someone who . . .	**Name**
is wearing his or her favorite color.	_____
likes to "dress up" in nice clothes.	_____
hates to wear high heels.	_____
wears black all the time.	_____
hates to wear a tie.	_____
always wears a belt.	_____
usually wears jewelry (rings, bracelets, etc.).	_____
likes to wear leather jackets, skirts, or pants.	_____
is wearing interesting socks.	_____
loves to shop for clothes.	_____

A **Are you wearing your favorite color?**
B **Yes. I'm wearing a purple shirt, and purple is my favorite color.**

Pair work Where in the world can you do these things? Think of a different country or city for each activity.

Where can you . . .	Name of country or city
ride an elephant or a camel?	Thailand, Egypt
climb a very high mountain?	
go snorkeling on a coral reef?	
go on a safari?	
visit several islands?	
take photos of amazing landscapes?	
see buildings over 500 years old?	
see a pyramid?	
go on a tour of a palace or a castle?	
take a cable car ride?	
hear traditional music?	
see traditional dance?	
eat very spicy food?	

A **Where can you ride an elephant?**
B **Let me think. I think you can ride elephants in Thailand.**
A **Right. Where can you ride a camel?**

1 Student A: Imagine that this is how your apartment looked this morning. What did you do yesterday? Study the picture for one minute, and try to remember.

2 *Pair work* Now close your book, and make a list of the things you did yesterday. Tell your partner. How many things can you remember? Then change roles. Look at **Free talk 10B**, and check your partner's answers.

1. made a cake

You ***I made a cake.***
Your partner ***That's right. What else?***

Free talk 11   *Guess where I went on vacation.*

1 Choose a beautiful or exciting city or country, and imagine you went there on vacation. Think of answers to these questions.

- How did you get there? Did you fly? Did you take a train or bus? Did you drive?
- How long did the trip take?
- What time of year was it?
- What language did they speak there?

- What was the weather like?
- What did you do there?
- What kind of food did you eat?
- What kinds of clothes did you wear?
- What kinds of souvenirs did you buy?

2 *Group work* Try to guess where each person went on vacation. Ask questions like the ones above (but don't ask *Where did you go?*). How many questions do you need to ask before you guess the city or country?

A ***So, how did you get there? Did you fly?***
B ***No, I took a bus.***
C ***How long . . . ?***

1 Student B: Imagine that this is how your apartment looked this morning. What did you do yesterday? Study the picture for one minute, and try to remember.

2 *Pair work* Now close your book, and make a list of the things you did yesterday. Tell your partner. How many things can you remember? Then change roles. Look at **Free talk 10A**, and check your partner's answers.

1. had coffee

You *I had coffee.*
Your partner *That's right. What else?*

Class activity Ask your classmates questions about their eating habits. Write their names in the chart. Write down any interesting information.

Find someone who . . .	Name	Notes
didn't have breakfast this morning.	Paula	doesn't like to eat in the morning
needs to drink coffee every morning.		
had a delicious dinner last night.		
doesn't eat rice, bread, or pasta.		
eats a lot of garlic or hot pepper.		
can't eat shellfish.		
thinks chocolate is good for your health.		
was a picky eater as a child.		
is or was a vegetarian.		
would like to eat something right now!		

A *Paula, did you have breakfast this morning?*
B *No, I didn't. I don't like to eat in the morning.*

Self-study listening

Unit 1

A *Track 1* Listen to the conversation on page 8. Alicia starts a conversation with Adam in the park.

B *Track 2* Listen to the rest of their conversation. Make these sentences true for Alicia and Adam. Circle the correct words.

Alicia:
1. *"My name is Alicia, or **Ally** / **Alice**."*
2. *"My last name is **Jones** / **Smith**."*
3. *"My middle name is **Katherine** / **Alicia**."*

Adam:
4. *"My favorite name is **Alicia** / **Katherine**."*
5. *"I'm **on vacation** / **not on vacation**."*
6. *"I'm **a music** / **an English** student."*

Unit 2

A *Track 3* Listen to the conversation on page 18. Ming-wei and Sonia are in English class.

B *Track 4* Listen to the rest of the conversation. Who asks these questions? Check (✓) the name.

	Ming-wei	Sonia	The teacher
1. "Can I borrow your dictionary, please?"	☐	☐	☐
2. "Where's my bag?"	☐	☐	☐
3. "What's the answer, please?"	☐	☐	☐
4. "Can you repeat the question, please?"	☐	☐	☐
5. "How do you spell **calendar**?"	☐	☐	☐
6. "Where's my dictionary?"	☐	☐	☐

Unit 3

A *Track 5* Listen to the conversation on page 28. Mark and Eve talk about Eve's friend Natasha.

B *Track 6* Listen to the rest of their conversation. Complete the sentences with *is* or *isn't*.

1. Mark says Natasha's painting _____ *interesting*.
2. Natasha _____ *very outgoing*.
3. Eve _____ *22 years old*.
4. Natasha _____ *Eve's best friend*.
5. Eve's brother _____ *in a band with Natasha's boyfriend*.
6. Mark _____ *pleased about Natasha's boyfriend*.

Unit 4

A *Track 7* Listen to the conversation on page 38. Tina starts a conversation with Ray in a coffee shop.

B *Track 8* Listen to the rest of their conversation. Check (✓) true or false for each sentence.

	True	False
1. Tina really likes Laguna Beach.	☐	☐
2. Tina's house is near the beach.	☐	☐
3. Tina goes to the beach on Saturdays.	☐	☐
4. Ray plays on a softball team.	☐	☐
5. Tina plays softball at a park near the coffee shop.	☐	☐
6. Ten guys play on Tina's softball team.	☐	☐

Unit 5

A *Track 9* Listen to the conversation on page 48. Adam and Lori are talking after class.

B *Track 10* Listen to the rest of their conversation. Choose the right answer. Circle **a** or **b**.

1. What does Lori say about Fabio's?
   a. It's always very busy.
   b. The food is really bad.

2. When is Adam usually free?
   a. On Mondays and Thursdays.
   b. On Tuesdays and Wednesdays.

3. Why is Adam so busy?
   a. He goes out with his friends a lot.
   b. He has a job and goes to school.

4. Why does Adam say "Oh, no!" to Lori?
   a. He doesn't go out on Thursdays.
   b. He doesn't want dinner at Fabio's.

Unit 6

A *Track 11* Listen to the conversation on page 58. Jessica and Ben are hungry.

B *Track 12* Listen to the rest of their conversation. Circle the correct words.

1. There are some restaurants **at the mall** / **near the park**.
2. Ben **has** / **doesn't have** a lot of money today.
3. Ben doesn't want **French food** / **fast food** today.
4. **Ben's** / **Jessica's** neighborhood is thirty minutes away.
5. The French restaurant is **fifteen** / **fifty** minutes away.
6. Jessica wants a snack **before** / **after** lunch.
7. Jessica usually eats lunch at **noon** / **2:30**.

Unit 7

A *Track 13* Listen to the conversation on page 70. Kate, Ray, and Tina are at a barbecue.

B *Track 14* Listen to the rest of the conversation. Check (✓) true or false for each sentence.

	True	False
1. Ray hates the beach.	☐	☐
2. Ray is a good swimmer.	☐	☐
3. Ray is giving Tina swimming lessons.	☐	☐
4. Ray practices swimming every day.	☐	☐
5. People from the softball team are at the barbecue.	☐	☐

Unit 8

A *Track 15* Listen to the conversation on page 80. Sarah is talking to a clerk.

B *Track 16* Listen to the rest of the conversation. Circle the correct words.

1. Sarah wants to pay for the bracelet with **cash** / **a credit card**.
2. The necklaces cost **$70** / **$17**.
3. The clerk says the **necklaces** / **earrings** are beautiful on Sarah.
4. Sarah thinks the earrings **are cheap** / **are expensive**.
5. Sarah **needs** / **doesn't need** new earrings.
6. Sarah **uses** / **doesn't use** her credit card to pay for the jewelry.

Unit 9

A *Track 17* Listen to the conversation on page 90. Yuki and Adam are ordering food at a coffee shop.

B *Track 18* Listen to the rest of the conversation. Choose the right answer. Circle *a* or *b*.

1. What's a smoothie?
   a. It's a kind of drink.
   b. It's a kind of ice cream.

2. What are smoothies like?
   a. They're like coffee with milk.
   b. They're like a milk shake with fruit.

3. Why does Yuki call the server?
   a. She wants a smoothie.
   b. She wants some sprinkles.

4. What's the problem with the order?
   a. Yuki doesn't understand the server.
   b. The server doesn't understand Yuki.

Unit 10

A *Track 19* Listen to the conversation on page 102. Eve is telling Mark about her busy week.

B *Track 20* Listen to the rest of their conversation. Complete the sentences. Choose *a* or *b*.

1. Mark tells Eve that _____ .
   a. he didn't have a very busy week
   b. he did a lot this week

2. Eve is happy about Mark's new computer because _____ .
   a. she needs to borrow it
   b. Mark doesn't need to borrow her laptop now

3. Mark wrote his last history paper _____ .
   a. with his new computer
   b. with Eve's laptop

4. This week Mark _____ on his new computer.
   a. watched DVDs
   b. wrote a paper

Unit 11

A *Track 21* Listen to the conversation on page 112. Jessica and Ben are talking about their weekend.

B *Track 22* Listen to the rest of their conversation. Answer the questions.

1. Why did Ben have a party? _____
2. Why didn't Jessica go to his party? _____
3. Why didn't Jessica get Ben's e-mail? _____
4. What happened on Jessica's bike trip? _____
5. What does Jessica want to do for Ben's birthday? _____

Unit 12

A *Track 23* Listen to the conversation on page 122. Emily and Matt are talking about lunch.

B *Track 24* Now listen to Emily and Matt's conversation in the restaurant. They're ordering lunch. Complete the sentences. Choose *a* or *b*.

1. Matt thinks that the restaurant _____ .
   a. has really good food
   b. has really good service

2. **Gazpacho** is a Spanish soup with _____ .
   a. fish, potatoes, and rice
   b. tomatoes, cucumbers, and peppers

3. Emily orders _____ .
   a. a sandwich and coffee
   b. soup and a sandwich

4. Matt wants to start with _____ .
   a. cheesecake
   b. gazpacho

5. Emily is surprised because _____ .
   a. Matt said he didn't want a big meal
   b. Matt said he didn't want soup

6. Matt changes his order because _____ .
   a. he doesn't have much money
   b. he's not very hungry

Unit 1

Adam   Your name is Alicia?

Alicia   Yeah, Alicia or Ally.

Adam   Ally?

Alicia   Yeah. Ally Jones.

Adam   Alicia's a nice name.

Alicia   Thank you. Actually, it's my middle name. My first name is Katherine.

Adam   Really? How do you spell it? With a *K* or a *C*?

Alicia   With a *K*. It's K-A-T-H-E-R-I-N-E.

Adam   Huh. My friend's name is Katherine. It's my favorite name. So, you're on vacation. . . .

Alicia   Yeah. How about you?

Adam   Uh, I'm a student here, a music student.

Alicia   Oh, right! Sorry. Well, nice meeting you.

Adam   Yeah. So, enjoy the concert.

Alicia   Thanks! You too. Bye.

Unit 2

Ming-wei   Sonia . . . Sonia. . . . Excuse me, Sonia! Can I borrow your dictionary, please?

Sonia   Sure. It's, um. . . . Oh. Where is it?

Ming-wei   I think it's there – in your bag.

Sonia   Oh, thanks. . . . Uh, where's my bag?

Ming-wei   On the floor. Look – under your chair.

Sonia   Oh! OK. . . . Here.

Ming-wei   Thanks.

Sonia   Sure.

Ms. Larsen   Sonia? Sonia?

Sonia   Uh-oh. . . . Yes?

Ms. Larsen   What's the answer, please?

Sonia   I'm sorry. Can you, uh, repeat the question, please?

Ms. Larsen   Sure. How do you spell *calendar*?

Sonia   Oh! It's, um . . . C-A-L, um. . . . Sorry. I don't know. . . . Excuse me, Ming-wei, where's my dictionary?

Unit 3

Eve   What is it? It's a . . . I think it's a . . . hmm. I don't know. But I think it's amazing.

Mark   Yeah, it's, uh, yeah. It's really interesting. So, what's Natasha like?

Eve   She's really nice, but she's shy.

Mark   Really? She's, uh, really good-looking. How old is she?

Eve   Well, we're the same age. So she's 22.

Mark   Oh. So is she your best friend?

Eve   Um . . . no. We're friends because her boyfriend and my brother are in the same band.

Mark   Her . . . boyfriend. Huh.

Eve   Yeah. Her boyfriend's the singer. He's really good. And their new CD is out now. It's great!

Mark   Really? That's, um . . . nice, really nice.

Unit 4

Ray   So, how do you like Laguna Beach?

Tina   Oh, it's great. I live right near the beach.

Ray   Nice! . . . By the way, my name is Ray.

Tina   I'm Tina. Nice to meet you.

Ray   Yeah. So, do you go to the beach a lot?

Tina   Yeah, on the weekends. Well, I go every Sunday. I play softball on Saturdays. I'm on a team.

Ray   Really? I play softball, too.

Tina   Oh. Do you play on a team?

Ray   Well, not right now.

Tina   So, come and join our team. We play at a park near here.

Ray   Uh, do guys play on your team?

Tina   Sure they do. We have, uh, ten women and five guys.

Ray   Hmm. Interesting.

Tina   Well, we have practice every Saturday morning. Just come on Saturday, and see how you like it.

Unit 5

Lori   You work at Fabio's? Oh! Well, I mean, the service isn't really bad. . . . It's just always, um, very busy.

Adam   That's OK. We *are* really busy.

Lori   Is it a part-time job? I mean, how often do you work there?

Adam   Well, I work every night on the weekends. And sometimes on Tuesdays and Wednesdays.

Lori   Huh. So you don't have a lot of free time.

Adam   No, I don't. I usually just study, go to classes, go to work.

Lori   So what do you do for fun? Do you go out with friends?

Adam   Yeah. Sometimes, but not very often.

Lori   Oh, that's not good. . . . Listen, come and have dinner with me and my friend tonight. I mean, it's Thursday night, so you're free – right?

Adam   Yeah, I am.

Lori   Good. I know a great restaurant. It has really good food, and the service is, uh . . .

Adam   Terrible? You mean Fabio's? Oh, no!

Unit 6

Jessica   Well, there are some restaurants at the mall near here.

Ben   Are they cheap? Uh, I don't have a lot of money with me, so . . .

Jessica   You know, they're, like, fast-food places.

Ben   Hmm. I don't really want fast food today. Uh, what about my neighborhood? There are some cheap restaurants there.

Jessica   But, uh, your neighborhood is 30 minutes from here by bus. And I'm really hungry. It's 2:30, you know.

Ben   Right. So, how about, um. . . . Oh, I know! There's a little French restaurant near here. It's good and it's not expensive.

Jessica   OK. So let's go there. Is it far?

Ben   It's about 15 minutes from here.

Jessica   Fifteen minutes. Well, OK. But let's get a snack first. Look. There's a supermarket right over there.

Ben   OK. Boy, you *are* hungry!

Jessica   Well, I usually have lunch at noon.

Answer key

Unit 1   1. Ally   2. Jones   3. Alicia   4. Katherine   5. not on vacation   6. a music

Unit 2   1. Ming-wei   2. Sonia   3. The teacher   4. Sonia   5. The teacher   6. Sonia

Unit 3   1. is   2. isn't   3. is   4. isn't   5. is   6. isn't

Unit 4   1. True   2. True   3. False   4. False   5. True   6. False

Unit 5   1. a   2. a   3. b   4. b

Unit 6   1. at the mall   2. doesn't have   3. fast food   4. Ben's   5. fifteen   6. before   7. noon

Unit 7

Kate So, Ray, uh, do you scuba dive?

Ray No, I don't. I love the beach, but, uh, I don't swim.

Kate Huh. . . . Oh, here's Tina.

Tina Hi. Sorry about the phone call. So, what are you guys talking about?

Kate Ray's telling me that he doesn't swim.

Tina Oh, yeah, I know. . . . But he's taking lessons!

Ray Yeah. Tina's my teacher!

Tina Ray's not a bad student, but he hardly ever practices.

Ray Hey! We have a lesson every week!

Tina Oh, look! Some guys from the softball team are here.

Ray Really? Where?

Tina Look over there.

Ray Oh, yeah. Hi, you guys!

Tina Watch out! The pool. Oh, no! . . . Ray, are you OK?

Ray Oh, I'm just fine. But that's enough swimming for this week.

Unit 8

Clerk OK. This is a beautiful bracelet. Are you paying cash or with a credit card?

Sarah Um . . . let me see. Um, cash.

Clerk OK. Uh, what about you? Do *you* need anything? We have some new necklaces. They're only $70.

Sarah Oh, they're really nice, but $70 is a lot of money.

Clerk Well, what about these earrings? Oh, look – they're beautiful on you.

Sarah Hmm. They really are. How much are they?

Clerk Uh, they're $99.99. . . . No, wait. They're on sale. 50% off. Uh, they're $49.99.

Sarah The earrings are on sale? 50% off?

Clerk Uh-huh.

Sarah $49 isn't bad. And I need some new earrings.

Clerk They *are* very nice.

Sarah Um, well, OK. I guess I'll take them.

Clerk OK. So, that's $105.94. You're paying cash, right?

Sarah It's $105.94? Um, well, actually, I think I'll use a credit card.

Unit 9

Yuki What are they called again?

Adam Sprinkles. They're good on ice cream or on a smoothie.

Yuki On a what? A smooth–?

Adam A smoothie.

Yuki A smoothie? What's that? Is it a kind of drink?

Adam Yes, it is.

Yuki So, what's it like? Coffee or an ice-cream soda or . . . ?

Adam It's, um, kind of like a milk shake, but with fruit.

Yuki It sounds good.

Adam Yeah, and smoothies are good for you, too.

Yuki Do they have smoothies here?

Adam Uh, let me see. Um . . . yeah, they do.

Yuki Good. Uh, excuse me?

Server Yes. What can I do for you?

Yuki Can I change my order? I want to try a smoothie.

Server Sure. Blueberry? Banana? Orange? Raspberry?

Yuki Uh, blue . . . ? I'm sorry, can you repeat all that, please?

Unit 10

Eve Thanks. Yeah. I really want to work there. So, how about you? Did you have a busy week?

Mark Um. . . . Well, no. I mean, I didn't do a lot. But I bought a computer.

Eve You did? Thank goodness! Now you don't, uh need to borrow my laptop. You had it for two weeks last time.

Mark Yeah, sorry about that. I wrote my history paper on it. Thanks again, by the way.

Eve Sure. No problem. So how do you like your computer?

Mark Oh, it's really cool. I use it every day.

Eve Really? Are you using it for class? I mean, you're working on a paper now, right?

Mark Uh, yeah, I am. But I don't use it for class.

Eve So, what do you use it for?

Mark Um, actually, I just watched DVDs on it all week!

Unit 11

Jessica Wait, Ben. Did you say you had a party Saturday?

Ben Yeah. You didn't know about it? It was my birthday.

Jessica No, I didn't.

Ben Yeah, I invited you. I sent you an e-mail last week.

Jessica Are you sure? I didn't get it. . . . Oh, you know what? I didn't tell you. I have a new e-mail address. Oh, I'm sorry.

Ben Oh, that's OK. But I was, uh, kind of upset that you weren't there.

Jessica Really? I'm upset, too. And that bike trip was just awful. We got lost . . . for four hours!

Ben Oh, that's terrible.

Jessica Listen, let's go out for lunch together this week. I want to do something for your birthday.

Ben Oh, you don't have to do anything.

Jessica I know, but I want to.

Ben OK. That's really nice of you. Really. . . . So, anyway, I have to go now. I have a meeting at 9:30.

Jessica OK, see you later. And, uh, happy birthday!

Unit 12

Matt Hey, this is a great menu.

Emily Yeah, I know. So, do you want soup or . . . ?

Matt I don't know. There are a lot of interesting fish dishes.

Emily Well, they have some of that Spanish soup – *gazpacho*.

Matt What's it called? Gazpacho?

Emily Yeah. It's a kind of cold soup with tomatoes, cucumbers, and peppers. It comes with bread.

Matt Hmm. Well, it's not hot, but it sounds good.

Server So, can I take your order? What would you like?

Emily I'll have a chicken sandwich and some coffee.

Matt And I'd like to start with some gazpacho. . . . And then I want that fish with rice, some potato salad, uh, iced tea. . . . And some cheesecake.

Emily Hey, you said you didn't want a big meal.

Matt I did, but everything looks so good. . . . Wait, where's my wallet? . . . Oh, no. I left my wallet at the office! I only have five dollars in my pocket! . . . Excuse me, sir.

Server Yes. Would you like something else?

Matt Can I change my order? I'd just like some of that soup – gazpacho – with, uh, a lot of bread and butter.

Answer key

Unit 7 1. False 2. False 3. False 4. False 5. True

Unit 8 1. cash 2. $70 3. earrings 4. are cheap
5. needs 6. uses

Unit 9 1. a 2. b 3. a 4. a

Unit 10 1. a 2. b 3. b 4. a

Unit 11 1. It was his birthday. 2. She didn't get Ben's e-mail.
3. She has a new e-mail address. 4. She got lost for four hours.
5. She wants to go out for lunch with Ben.

Unit 12 1. a 2. b 3. a 4. b 5. a 6. a

Illustration credits

Laurie Conley: 16 *(bottom)*, 20 *(top)*, 68 *(top)*
Susan Gal: 10, 20 *(bottom)*, 30, 42 *(bottom)*, 52, 62, 74, 84, 106, 116, 126
Peter Hoey: 40, 63, 100, 101
Adam Hurwitz: FT-D
Kim Johnson: 9, 35, 39, 59, 71, 78 *(top)*, 109, 125
Marilena Perilli: x, 5, 6, 14, 16 *(top)*, 17, 31, 32, 57, 66, 78 *(bottom)*, 79, 96, 110, 120, 121, FT-A *(top)*, FT-E

Tony Persiani: 83
Andrew Shiff: 54
Andrew Vanderkarr: 4, 12, 13, 24, 25, 26, 27, 46, 68 *(bottom)*, 73, 86, 87, 99, 111, 118, 119
Dan Vasconcellos: 50, 128
Filip Yip: 15, 19, 56 *(middle)*, 81, FT-A *(bottom)*, FT-B, FT-G, FT-H

Photography credits

2, 3, 8, 18, 19, 28, 29, 34, 38, 39, 48, 49, 58, 59, 70, 71, 80, 81, 102, 112, 122, SSL1, SSL2, SSL3 ©Frank Veronsky

1 *(clockwise from top left)* ©Alamy; ©Corbis; ©Alamy

4 *(left to right)* ©First Light; ©First Light; ©Getty Images

6 *(left to right)* ©Orlando Marques/First Light; ©Chabruken/Getty Images; ©Alamy

9 *(top, left to right)* ©Getty Images; ©Anthony Redpath/Corbis

11 *(clockwise from top right)* ©Jose Luis Pelaez Inc./Corbis; ©Michael Goldman/First Light; ©Jose Luis Pelaez Inc./Corbis; ©Gabe Palmer/Corbis

15 ©Stephen Ogilvy

21 *(clockwise from top right)* ©Joe McBride/Corbis; ©Corbis; ©RJ/Big Pictures USA/Newscom; ©Ariel Skelley/Corbis

22 *(top, left to right)* ©Corbis; ©Kevin Winter/ImageDirect/Getty Images; ©Chris Hondros/Newsmakers/Getty Images; *(bottom)* ©Frank Veronsky

23 *(top to bottom)* ©Corbis; ©Tim Mosenfelder/Getty Images; ©Rufus F. Folkks/Corbis; ©Warner Bros. Television/Getty Images; ©Leonard Ortiz/Newscom

33 *(clockwise from top right)* ©Getty Images; ©Jose Luis Pelaez Inc./Corbis; ©Tom Stewart/Corbis; ©Getty Images

36 *(clockwise from top right)* ©Rob Gage/Getty Images; ©Mike Okoniewski/The Image Works; ©Frank Herholdt/Getty Images; ©Getty Images; ©Bob Handelman/Getty Images; ©Brian Pieters/Masterfile

37 ©Adrian Weinbrecht/Getty Images

43 *(clockwise from top right)* ©Guy Cali/Alamy; ©GDT/Getty Images; ©Getty Images; ©Ed Bock/Corbis

44 *(counterclockwise from top right)* ©Getty Images; ©Getty Images; ©PhotoDisc; ©Corbis; ©PhotoDisc

45 ©Rubberball/First Light

47 *(clockwise from top left)* ©Warner Bros./Courtesy Neal Peters Collection; ©Everett Collection; ©Reuters/Corbis; ©Everett Collection; ©ABC/Neal Peters Collection; ©NBC/Neal Peters Collection; ©Everett Collection; ©Getty Images

51 *(left to right)* ©Mark Leibowitz/Masterfile; ©Getty Images

53 *(clockwise from top right)* ©Kathey Willens/AP; ©Bruce Burkhardt/Corbis; ©Albert Normandin/Masterfile; ©Andre Jenny/Alamy

54 *(left to right)* ©Alamy; ©Picture Quest; ©Corbis

61 *(bottom left)* ©Getty Images; *(bottom right)* ©Susan Werner/Getty Images

65 *(clockwise from top left)* ©Bonnie Kamin/PhotoEdit; ©Getty Images; ©Mike Powell/Getty Images

67 *(top to bottom)* ©Sakis Papadopoulos/Getty Images; ©Laureen March/Corbis

69 ©Reuters/Corbis

72 ©Jim Cummins/Getty Images

75 *(left to right)* ©Getty Images; ©Picture Quest; ©Punchstock; ©Tim Kiusalaas/Masterfile

76 ©Stephen Ogilvy

82 *(clockwise from left)* ©Dennis Marsico/Corbis; ©Robert Holmes/Corbis; ©Andrea Pistolesi/Getty Images

85 *(clockwise from top right)* ©Jeremy Horner/Corbis; ©Cosmo Condina/Getty Images; ©Corbis; ©Corbis; ©Corbis

86 *(left, top to bottom)* ©Corbis; ©Corbis; ©Gail Mooney/Corbis; ©Corbis; ©Comstock

89 *(clockwise from top left)* ©Mark Rightmire/Newscom; ©James Baigrie/Getty Images; ©Studio SatoPhotonica; ©Zav Mansourian/Getty Images; ©Richard T. Nowitz/Corbis; ©Steven Needham/Envision; ©Greg Elms/Lonelyplanet; ©Brian Hagiwara/Getty Images

90 *(top)* ©Frank Veronsky; *(bottom, left to right)* ©Catherine Karnow/Corbis; ©Spencer Grant/PhotoEdit; ©Bohemian Nomad Picturemakers/Corbis; ©Stephen Ogilvy

91 *(top, left to right)* ©Frank Veronsky; ©Adventure House; *(bottom, left to right)* ©Getty Images; ©Pawel Libera/Alamy; ©Doug Scott/age Fotostock; ©Richard T. Nowitz/Photo Researchers

92 *(top to bottom)* ©Michael Duerinck/Image State; ©B.S.P.I./Corbis; ©Corbis; ©Ashley Simmons/Alamy

93 ©Pablo Corral Vega/Corbis

97 *(clockwise from top right)* ©Picture Quest; ©Michael Cogliantry/Getty Images; ©Cary Wolinsky/Stock Boston Inc./Picture Quest; ©Marc Romanelli/Getty Images

98 *(clockwise from top left)* ©Jonathan Cavendish/Corbis; ©Nicolas Russell/Getty Images; ©Getty Images; ©Picture Quest; ©S. Hammid/Masterfile; ©Shaz/Retna; ©Ken Reid/Getty Images

103 *(top to bottom)* ©Frank Veronsky; ©Corbis

107 *(clockwise from top left)* ©Getty Images; ©Punchstock; ©Getty Images

108 *(clockwise from top left)* ©Bonnie Kamin/PhotoEdit; ©William A. Bake/Corbis; ©Chabruken/Getty Images; ©Getty Images

113 *(both top)* ©Frank Veronsky; *(bottom)* ©Alamy

114 ©Getty Images

115 *(left to right)* ©Alamy; ©Digital Vision

117 *(clockwise from left)* ©Comstock; ©Getty Images; ©Corbis; ©Steve Cohen/Getty Images; ©Penina/Getty Images; ©Getty Images

120 ©Stephen Ogilvy

123 *(top to bottom)* ©Frank Veronsky; ©Frank Veronsky; ©Punchstock; ©Punchstock; ©Digital Vision

127 *(top to bottom)* ©Alamy; ©Getty Images

FT-A *(both)* ©Reuters/Corbis

FT-B ©Reuters/Corbis

FT-C *(top left)* ©Steve McAlister/Getty Images

FT-F *(clockwise from top left)* ©age Fotostock; ©Galen Rowell/Corbis; ©IT Stock/Punchstock; ©Getty Images; ©Chad Ehlers/Getty Images; ©Reuters Photo Archive/Newscom; ©Hollenbeck Photography/Image State; ©Brian Hagiwara/Getty Images

FT-G ©Jerome Tisne/Getty Images

Text credits

Every effort has been made to trace the owners of copyrighted material in this publication. We would be grateful to hear from anyone who recognizes his or her copyrighted material and who is unacknowledged. We will be pleased to make the necessary corrections in future editions.

Component List for *Touchstone* 1

Title	中文书名	ISBN
Teacher's Edition 1 (with Audio CD)	剑桥标准英语教程1（教师用书）（附光盘1张）	978-7-5619-2656-7
Student's Book 1 (with Audio CD/CD-ROM)	剑桥标准英语教程1（学生用书）（附光盘1张）	978-7-5619-2649-9
Student's Book 1A (with Audio CD/CD-ROM)	剑桥标准英语教程1A（学生用书）（附光盘1张）	978-7-5619-2650-5
Student's Book 1B (with Audio CD/CD-ROM)	剑桥标准英语教程1B（学生用书）（附光盘1张）	978-7-5619-2651-2
Workbook 1	剑桥标准英语教程1（练习册）	978-7-5619-2652-9
Workbook 1A	剑桥标准英语教程1A（练习册）	978-7-5619-2653-6
Workbook 1B	剑桥标准英语教程1B（练习册）	978-7-5619-2654-3
Video Resource Book 1	剑桥标准英语教程1（Video活动用书）	978-7-5619-2655-0
Class Audio CDs 1	剑桥标准英语教程1（Class Audio CDs）	978-7-88703-911-8
Video 1	剑桥标准英语教程1（Video）	978-7-88703-912-5

新东方 NEWORIENTAL

Real
剑桥实境英语系列
实用英语权威教程，真实再现听·说·读·写日常情景！

Listening & Speaking

剑桥实境英语 **1**
Real
听说 Listening & Speaking with answers
Miles Craven
编号: 90608
定价: 35.00元

剑桥实境英语 **2**
Real
听说 Listening & Speaking with answers
Sally Logan and Craig Thaine
编号: 90609
定价: 35.00元

剑桥实境英语 **3**
Real
听说 Listening & Speaking with answers
Miles Craven
编号: 90610
定价: 35.00元

剑桥实境英语 **4**
Real
听说 Listening & Speaking with answers
Miles Craven
编号: 90611
定价: 35.00元

Reading

剑桥实境英语 **1**
Real
阅读 Reading
Liz Driscoll
编号: 90616
定价: 30.00元

剑桥实境英语 **2**
Real
阅读 Reading
Liz Driscoll
编号: 90617
定价: 30.00元

剑桥实境英语 **3**
Real
阅读 Reading
Liz Driscoll
编号: 90618
定价: 30.00元

剑桥实境英语 **4**
Real
阅读 Reading
Liz Driscoll
编号: 90619
定价: 30.00元

Writing

剑桥实境英语 **1**
Real
写作 Writing
Graham Palmer
编号: 90612
定价: 35.00元

剑桥实境英语 **2**
Real
写作 Writing
Graham Palmer
编号: 90613
定价: 35.00元

剑桥实境英语 **3**
Real
写作 Writing
Roger Gower
编号: 90614
定价: 35.00元

剑桥实境英语 **4**
Real
写作 Writing
Simon Haines
编号: 90615
定价: 35.00元

购书方式

1. 新东方大愚连锁书店购买；详情登录http://www.dogwood.com.cn/dysd
2. 新东方图书销售网点购买；详情登录http://www.dogwood.com.cn/agentlist.asp
3. 新东方大愚图书官方网店购买；详情登录http://xdfdy.taobao.com
4. 新东方书友会邮购：北京市海淀区海淀东三街2号欧美汇大厦19层 / 邮编：100080 / 收款人：书友会 / 咨询电话：010-62605127
   注意：汇款金额 = 书价总额＋3元挂号费
   请在汇款单的附言栏写清书的编号、册数；请写清您的地址、邮编、姓名。
   为便于及时联系，保证书籍安全到达，请务必在汇款单附言栏中注明您的联系电话。

浏览更多精彩图书，请登录大愚图书网：www.dogwood.com.cn

商务英语"标杆"之作
BEC备考最新权威教程

● 剑桥资深专家编著 BEC考试真题演练 ●

剑桥标准商务英语教程（初级）	剑桥标准商务英语教程（中级）	剑桥标准商务英语教程（高级）
学生用书(附赠自学手册和MP3光盘)	学生用书(附赠自学手册和MP3光盘)	学生用书(附赠自学手册和MP3光盘)
●【ISBN】978-7-5605-2972-1	●【ISBN】978-7-5605-2973-8	●【ISBN】978-7-5605-2974-5
● 编 号：81011	● 编 号：81012	● 编 号：81013
● 定 价：68 元	● 定 价：68 元	● 定 价：68 元
教师用书	教师用书	教师用书
●【ISBN】978-7-5605-2975-2	●【ISBN】978-7-5605-2976-9	●【ISBN】978-7-5605-2977-6
● 编 号：81008	● 编 号：81009	● 编 号：81010
● 定 价：28 元	● 定 价：28 元	● 定 价：28 元

 购书方式

1. 新东方大愚连锁书店购买：详情登录http://www.dogwood.com.cn/dysd
2. 新东方图书销售网点购买：详情登录http://www.dogwood.com.cn/agentlist.asp
3. 新东方大愚图书官方网店购买：详情登录http://xdfdy.taobao.com
4. 新东方书友会邮购：北京市海淀区海淀东三街2号欧美汇大厦19层／邮编：100080／收款人：书友会／咨询电话：010-62605127
   注意：汇款金额 = 书价总额＋3元挂号费
   请在汇款单的附言栏写清书的编号、册数；请写清您的地址、邮编、姓名。
   为便于及时联系，保证书籍安全到达，请务必在汇款单附言栏中注明您的联系电话。

浏览更多精彩图书，请登录大愚图书网：www.dogwood.com.cn